A TI Graphics Calculator Approach to
Calculus

John T. Gresser

PRENTICE HALL, Upper Saddle River, NJ 07458

Executive Editor: George Lobell
Editorial Assistant: Gale A. Epps
Special Projects Manager: Barbara A. Murray
Production Editor: Wendy Rivers
Supplement Cover Manager: Paul Gourhan
Supplement Cover Designer: Liz Nemeth
Manufacturing Buyer: Alan Fischer

© 1999 by Prentice Hall
Upper Saddle River, NJ 07458

All rights reserved. No part of this book may be
reproduced, in any form or by any means,
without permission in writing from the publisher.

Printed in the United States of America

10 9 8 7 6 5 4 3 2 1

ISBN 0-13-020020-4

Prentice-Hall International (UK) Limited, London
Prentice-Hall of Australia Pty. Limited, Sydney
Prentice-Hall Canada, Inc., Toronto
Prentice-Hall Hispanoamericana, S.A., Mexico
Prentice-Hall of India Private Limited, New Delhi
Pearson Education Asia Pte. Ltd., Singapore
Prentice-Hall of Japan, Inc., Tokyo
Editora Prentice-Hall do Brazil, Ltda., Rio de Janeiro

To my wife and friend Pamela,
for encouraging my work

Contents

	Preface	vii
1	**Functions, Limits, and Calculators**	**1**
	1.1 The Graphing Calculator Philosophy	1
	1.2 Getting Started on the TI-83, and TI-86	3
	1.3 Functions in Mathematics	3
	1.4 Applications	8
	1.5 Limits of Functions	10
	1.6 Exercise Set	14
2	**Differentiation**	**21**
	2.1 The Definition	21
	2.2 Computing Derivatives	24
	2.3 Exercise Set	27
3	**Applications of Differentiation**	**31**
	3.1 Graphing Polynomials	31
	3.2 Graphing Functions	33
	3.3 The Mean Value Theorem	35
	3.4 Newton's Method	36
	3.5 Applications	39
	3.6 Exercise Set	42
4	**Integration**	**45**
	4.1 Area and the Definite Integral	45
	4.2 The Fundamental Theorem of Calculus	51
	4.3 Some Theorems of Integration	53
	4.4 Numerical Integration	55
	4.5 Exercise Set	58
5	**Applications of Integration**	**61**
	5.1 Area and Volume	61
	5.2 Mass and Center of Mass	64
	5.3 Miscellaneous Problems	68
	5.4 Exercise Set	69

6 The Transcendental Functions — 73
- 6.1 Inverse Functions 73
- 6.2 Logarithmic and Exponential Functions 79
- 6.3 The Inverse Trigonometric Functions 85
- 6.4 Exponential Growth and Decay 90
- 6.5 The Hyperbolic Functions and Their Inverses 93
- 6.6 Indeterminate Forms and L'Hopital's Rule 94
- 6.7 Exercise Set 96

7 Sequences and Series — 101
- 7.1 Sequences 102
- 7.2 Series 109
- 7.3 Positive Term Series 113
- 7.4 Alternating Series and Absolute Convergence 117
- 7.5 Power Series 121
- 7.6 Exercise Set 126

8 Plane Curves and Polar Coordinates — 131
- 8.1 Parametrically Defined Planar Curves 131
- 8.2 Polar Coordinates 138
- 8.3 Exercise Set 143

9 Vectors and Analytic Geometry — 151
- 9.1 Addition and Scalar Multiplication 153
- 9.2 The Dot and Cross Products 157
- 9.3 Lines and Planes 162
- 9.4 Exercise Set 167

10 Vector Valued Functions — 173
- 10.1 Basic Calculations 173
- 10.2 Motion in Space 179
- 10.3 Geometry of Curves 184
- 10.4 Exercise Set 188

Epilogue — 193

Index — 195

Preface

The last decade of the twentieth century has been an exciting and interesting time for the study of mathematics. At the dawn of the new millennium, mathematics is undergoing a technological change which will have a profound effect on the way mathematics is studied, understood and used in the future. Elaborate and powerful software applications on desk top computers are taking mathematics in one direction, and powerful hand-held graphics calculators are taking mathematics in another equally interesting direction. Mathematics is indeed fortunate to have benefitted from the efforts of corporations like Texas Instruments, which has produced the TI-82, TI-83, TI-85, TI-86 graphics calculators used in this book.

Calculus and Calculators

This manual is not meant to be a self-contained calculus text, but rather a supplement to a regular calculus text. Topics are covered in a way which is meant to parallel the sequence of topics in a fairly typical calculus text.

Additionally, this book is about calculus, not about the graphics calculators we will be using. Each calculator in the Texas Instruments series comes with a **Guidebook** which should used as a reference and as a source book for more detailed explanations about how to use the calculator.

Actually, as you will see, a fair amount is written about how to use the calculator in the early chapters, but the emphasis quickly shifts back to the study of mathematics, where it will remain for the rest of the book. As involved as a graphics calculator may be, it is also straight forward enough that it will come to be understood quite naturally, while attention is paid primarily to calculus. New menus, and commands on the calculator are explored in the context of pursuing some mathematical idea.

Chapter 1 is basically review in nature, except for the section on limits, but it is also an opportunity to learn some calculator fundamentals, and it should be covered by everyone with no previous experience with the graphics calculator being used. This chapter also provides an opportunity to get used to the practice of thinking about mathematics in a more technological way. It takes some time to adjust to this merger of mathematics and technology. Chapter 1 is meant to nudge us gently in this direction.

Some of the problems at the end of each section are meant to provide routine experience in using a calculator. **Routine problems from your main calculus text can also be used as calculator practice problems.** You can, with such problems, **look up their answers in the appendix of your main text** to verify that your calculator is being used correctly.

Problems are, however, also designed to foster an attitude of skepticism and experimentation. The importance of adopting a skeptical scientific attitude has already been discussed. Answers are, by design, not provided in the back. **How do you know** that you have the correct answer? **Is there another** solution to that equation which is being solved? Is there an interesting feature to a graph under consideration which is **too small to be seen** in a window, or which occurs outside the window being used? Problems in this manual sometimes create **unexpected, incomplete, or occasionally wrong answers**. In the first few problem sets, you will usually be alerted to look for unexpected results, but eventually, such warnings will not be supplied. **Whenever possible or appropriate, you should supply evidence that your answer is correct.** One of the goals of this manual is to develop a good, skeptical, scientific attitude.

Calculators can be used to gain a deeper understanding of calculus, by focusing complete attention on an issue of calculus rather than on a computation. Some of the problems are designed with this in mind.

Without question, calculators can be used to enhance problem solving skills. Just imagine the creative freedom that you will have when you can focus all of your energy on ideas rather than computations. Calculators make more substantial and interesting mathematical problems accessible, and this may be its most important contribution to mathematical education. One should strive to get as much experience as possible in doing problems of this sort. Problems designed to improved problem solving skills are included in the exercise sets along with all of the other problems.

Many of the exercise sets have additional problems, labeled **"Projects,"** which are somewhat more involved. They range in difficulty from being just longer and more interesting versions of ordinary problems, to being quite difficult. They should be accessible without outside background reading. These problems are designed to enhance problem solving skills, by making use of not only current topics under discussion, but, occasionally, a wide variety of previously discussed topics as well. At least some of them are presented in a playful way, and are meant to be enjoyed, as well as to be instructive.

These projects, however, should be tackled with some discretion as well. Using a calculator is an interesting, but very different way of doing mathematics, and it takes some time to get accustomed to it and to take advantage of all the opportunities it presents to the user. When you are ready to take on the issue of "putting it all together," you are encouraged to work on the projects that appear after the exercise sets at the ends of the chapters.

How to Read this Manual

There are four graphics calculators from Texas Instruments that could be used with this manual, but only two will actually be discussed. The TI-82 has been discontinued, but it is probably still widely available. If you are using a TI-82, it is similar enough to the TI-83 that it can be used with this manual by following the instructions designed for the TI-83. The TI-83 is, however, an improved version of the TI-82, and occasional adjustments will be necessary. The TI-85 and TI-86 are also quite similar, and we shall only discuss the TI-86.

The TI-83 and TI-86, however, are actually quite different. A manual that covers both of them would be difficult to read if we constantly alternated our discussion from one instrument to the other. To avoid this annoyance, we adopt the policy of writing about just one of them—we choose the TI-83—as the "calculator of choice." The TI-86 will always be referred to parenthetically.

If you are using a Ti-85/86, plan to read all of the material in the manual. Some of it may contain TI-82/83 specific material, but it will also contain mathematical ideas, and more general technical material that will apply to all calculators. When necessary, (★86★ brief instructions exclusively for the TI-86 will be enclosed parenthetically in this way ★86★). More expansive material pertaining exclusively to the TI-86 will be separated from the rest of the text by using the following two slightly different separating lines, the first to start the discussion, and the second to signify that it is over.

─────────
★★★★86★★★★
This space between separator lines is reserved exclusively for TI-86 discussion.
★★★★86★★★★
If you are using a TI-82/83, you may disregard the material between these separators.

Each key on your calculator has a white, gold or green (★86★ blue ★86★) function, and we will use slightly different symbols to denote each of the three different key colors. For example, the white tangent key is denoted by $\boxed{\text{TAN}}$, the gold π key by $[\![\pi]\!]$, and the green or blue "A" key by $(\!(\text{A})\!)$.

The end of each example is marked with a ■ symbol. Bold face text is used to emphasize certain ideas. It is also used to set off some menu items and commands on the calculator that might otherwise be lost in the main text. Otherwise, menu items and commands on the calculator are given the same "look" in the text as they have on the calculator.

Explanations of new commands are given just once, when they are first introduced, so if chapters or sections are skipped, it would help to skim over the material that was skipped, looking for and reading the material on the introduction of new commands. Since new commands appear in bold face type when they are first introduced, this should be relatively painless.

This material can be read and understood without a calculator, but surely the best way to use it is to work on a calculator at the same time. Enter the same (or similar) expressions on your calculator as you read the manual, so that you can experience the results first hand.

Some Closing Thoughts

No attempt has been made to make this manual a complete study of the Texas Instruments series of graphics calculators. **You are encouraged to explore on your own, and to seek help frequently in the** Guidebook **that accompanies your calculator.**

Exercises should be presented in an organized, thoughtful and readable manner. Projects, in particular, should be done with great care paid to presentation. Think of a project as a term paper, or as a report which is going to your supervisor at work. The same matters of presentation should play a role in creating a paper on any subject—including mathematics.

A pencil and paper strategy session can be a useful way to start a calculator work session. It should be a strategy session, however, and not a complete pencil and paper solution, unless, of course, such a solution is desired. Try to get the calculator to do as much of the work as possible.

And finally, above all, remember that **all technology is created by human beings**. Human beings make mistakes. Calculators (probably) make mistakes, infrequently, perhaps, but mistakes nevertheless. Human beings are, of course, a more common source of mistakes. A small mistake in an input statement can have enormous consequences. Even without an input line mistake, a computation can be misinterpreted, or used improperly by us with grave consequences.

Work with technology should be a partnership between a human being and a machine, not a thoughtless ride on a machine. The way to avoid wrong answers is to know mathematics well enough to see the warning signs when mistakes have been made, or when technology is misbehaving. **Make it a practice to verify that answers are correct, especially if an answer "looks" doubtful**. Frequently, answers can be verified very quickly.

In order to keep the main story line simple, brief, and readable, this manual will not always participate in this practice of verifying answers. Make it a habit, when reading this manual, and when doing your own work to **check your calculator's performance**.

Acknowledgments

I am indebted to the staff of Prentice Hall for its help in preparing this book, and to George Lobell, mathematics editor, for his encouragement. Finally, special thanks go to all of my students, past and present, who were or are a part of my technology based calculus program. Their cooperation and sense of humor kept this project moving forward.

─── ★ ★ ★ ───

Solving problems with technology, especially more involved problems, can be a rich and rewarding mathematical experience. I hope that this manual and its problems meet with your approval. My e-mail address is included below, because good text books are a community effort, and your comments and suggestions for improvements would be greatly appreciated.

John T. Gresser
Department of Mathematics and Statistics
Bowling Green State University
Bowling Green, OH 43403
jgresse@bgnet.bgsu.edu

Chapter 1

Functions, Limits, and Calculators

Calculus depends quite strongly on a variety of topics that are covered in courses leading up to this subject. It draws its strength from algebra and trigonometry, from polynomials, exponential and logarithmic functions, from function theory in general, from applications, and many other topics. Central to calculus, and to all of mathematics, is the idea of a function, and so it is appropriate to begin our study by looking at functions and their applications from a graphics calculator point of view.

This chapter is also an opportunity to get acquainted with the graphing calculator that we will be using throughout our study of calculus. Before we do anything else, it is essential that we study some of the basic ideas of working with a graphing calculator.

After we get acquainted with our calculator, we will use it to gain some insight into the basic idea of a limit. Just as the idea of a function is central to all of mathematics, the idea of a limit is central to all of calculus.

1.1 The Graphing Calculator Philosophy

Learning how to use a graphics calculator to do mathematics is more than just learning how to use all of its buttons, and menus. Technology also gives us another way of thinking about problem solving. **A problem might not be intrinsically a graphical problem, but if it can be posed in a graphical way, it increases our problem solving power enormously.** When our eyes can be used in the problem solving process, our intuition is much more engaged, and mathematics seems much less abstract and intractable.

For example, solving a given equation of the form $expr_1(x) = expr_2(x)$ for x, can be thought of as a very abstract problem. Certainly it is not, on the face of it, a graphical problem. Does a solution exist? How many solutions are there? Such questions can be difficult or impossible to answer in the abstract.

With a graphing calculator, however, we can easily graph the function $f(x) = expr_1(x) - expr_2(x)$, and the solutions to the above equation simply become points where this graph crosses the x-axis. (These are the points where $f(x) = 0$.) Equivalently, we could graph the curves $y = expr_1(x)$, and $y = expr_2(x)$. Then the sought after solutions to the equation become the x-coordinates of the points of intersection of the two curves.

There are many other mathematical problems which might originate is a more abstract non-graphical way, and which could be stated in a more beneficial way by looking at them graphically. Make it a habit to look for a visual way to analyze the problems you encounter in mathematics. Initially, this may take some effort. Until you get used to a graphical point of view, it may not occur to you to look in this direction unless you make a point of it.

There are analytical and numerical ways to solve mathematical problems as well, and we are not suggesting that a graphical approach should be the only approach used. It may not even be appropriate to use graphical evidence in the solution of a particular problem. However, if a graphics calculator can be used to give us insight, then it can be an effective tool, even if we must produce an analytic solution to a problem. A graphics calculator is simply another tool that can be used in the problem solving process. In this text, you will be encouraged to use your calculator frequently, but the temptation to depend on it too much, or when it is not appropriate should be resisted.

Technology also offers us more of an opportunity to experiment, to play "what if" games. Confronted with an intractable problem, it might help to solve a similar, but simpler problem. If there are arbitrary constants involved, some insight might be gained by giving them special values. We can make conjectures and use a calculator to test whether the conjecture seems to be true or false. In a pencil and paper world there is a heavy price to pay for experimentation—we have to do the resulting calculations. Technology, on the other hand, often makes experimentation easier to perform.

Mathematicians, indeed, all scientists, are naturally skeptics. Their first reaction to any claim is to ask for a proof, and then they are inclined to look for mistakes and counter examples when they study the work of their colleagues. It is exceedingly important that students of science and mathematics acquire this attitude of skepticism. **Is the answer reasonable? Is it correct? Are there other answers?** These are questions we should always ask, and they should still be asked, even when technology is used to solve problems. A small mistake in an input statement can have enormous consequences. Even without an input line mistake, a computation can be misinterpreted, or used improperly by us with grave consequences. Work with a graphics calculator should be a partnership between a human being and a machine, not a thoughtless ride on a machine. The way to avoid wrong answers is to know mathematics well enough to see the warning signs when mistakes have been made, or when technology is misbehaving. Make it a practice to verify that answers are correct, especially if an answer "looks" doubtful. An alert mind and a skeptical attitude are a critical part of using technology in a successful way.

1.2 Getting Started on the TI-83, and TI-86

For the most part, we will learn to use our graphing calculator by using it to do mathematics. There are, however, some fundamental operations we should become familiar with before we do anything else. The manual that came with your calculator is an excellent source for this preliminary work and we will depend on it to give us this jump start.

If you are using a TI-83, read the section entitled **Getting Started: Do This First**, on pages 1-17 of the TI-83 Guide Book. (★86★ If you are using a TI-86, you should read the section entitled **T1-86 Quick Start** on pages 1-14 of the TI-86 Guidebook. ★86★) It may help to turn on your calculator and reproduce the activity on your own calculator. As we continue, we will assume a passing familiarity with the items on these pages from your calculator's manual.

The TI-82/83 and TI-85/86 are actually quite different. A manual that covers all of them would be difficult to read if we constantly alternated our discussion from one instrument to the other. To avoid this annoyance, this manual has adopted certain conventions which are described in the Preface under the section entitled "How to Read this Manual". You are encouraged to read this section before continuing.

To avoid unwanted settings that you may have inherited with your new calculator, it is suggested that you reset it to its default factory settings. With your calculator turned on, press [2ND] [MEM] [5] [2] [2] [ENTER] (★86★ press [2ND] [MEM] [F3] [F3] [F4] [ENTER] ★86★).

1.3 Functions in Mathematics

A real valued function f defined on a set D of real numbers is a rule which assigns a unique real number to each x in D. **We define functions on our calculator by using the Y= menu.** Press [Y=] move the cursor to one of the labels Y1, Y2, ..., Y0 (Y0=Y10), and enter a formula in the expected way. Ten different functions can be stored either **actively or inactively** in this way. **Graphs and tables are not shown** for functions which are stored **inactively**. If the (=) **symbol** for the function is **highlighted** (the default mode), then the function is in an **active mode**. To change a function from active to inactive or vice-versa, move the cursor to the equal sign and press [ENTER]. Notice how the highlighting on the equal sign switches on or off. Turning functions on or off in this manner can be a real convenience.

The top row of buttons on the TI-83 is central to all of the calculator based activity in this manual. Read the Getting Started section in the Guide Book for this important information.

★★★★ 86 ★★★★

On the TI-86 press [GRAPH] to enter an elaborate system of menus, submenus, and commands used to define, and graph functions. Generally speaking, in all of the TI-86 menus, the F-keys select items in the bottom menu bar. When there are two menu bars (there usually is) the M-keys select items in the top menu bar. To create a function, open the $y(x) =$ menu. From the home screen this is done by pressing

[GRAPH] and then [F1]. Functions $y1, y2, \ldots, y99$ can be entered in the expected way. To change a function from an active to an inactive mode or vice-versa, press [F5] while you are in the $y(x) =$ menu.

This system of menus, submenus, and commands is a critical part of the TI-86 calculator. In addition, many (though not all) of the items in the [GRAPH] menu are central to all of the calculator based activity in this manual. Read the TI-86 Quick Start section in the Guide Book for this important information.

$\star\star\star\star 86 \star\star\star\star$

Example 1.1 *Graph the function $f(x) = \sin^3(x)/(x^2 + x + 3)$ on the interval $[-2\pi, 2\pi]$. Adjust the vertical scale so that the graph fits nicely in the graphing window. Use the zoom feature of your calculator to find the coordinates of the high point and the low point on the graph with an accuracy to at least 3 digits. Explain why the desired level of accuracy has been achieved.*

From the home screen, press [Y=] and enter the function. If any expressions appear in this screen from previous work they should be removed by moving the cursor to that location, and pressing [CLEAR]. Alternately, unwanted expressions could simply be deactivated as mentioned above.

The expression $\sin^3(x) = (sin(x))^3$ can be entered in the form $\sin(X)\wedge 3$ or in the form $(\sin(X))\wedge 3$. The extra parentheses are unnecessary ($\star 86\star$ Surprisingly, the extra parentheses are necessary on this calculator. More on this later. $\star 86\star$) When there is any doubt, it is best to err on the side of extra parentheses.

Press [WINDOW] and enter $X_{min} = -2\pi$ by moving the cursor to that location and pressing [CLEAR] [(−)] [2] [2ND] [π]. **Use the** [(−)] **key and not the** [−] **key**. In the same manner, enter $X_{max} = 2\pi$. Entries for Y_{min}, and Y_{max} can be entered in a similar manner, if it is evident what values to choose. Frequently, however, it will not be clear. To have the calculator select a "best fit", press [ZOOM], move the cursor down to **ZoomFit**, the tenth item (menu item 0), and press [ENTER]. Alternately, press [ZOOM] [0]. Using either approach, the graph should now appear on screen.

To find the high point on the interval $[-2\pi, 2\pi]$ press [TRACE] and use the ▶ and ◀ keys to move the trace cursor as close as possible to the highpoint. Press [ZOOM] [2] [ENTER] to "Zoom In" on the point. Equivalently, press [ZOOM] ▼ to highlight the second menu item (Zoom In) and press [ENTER]. Repeat this procedure (several times if necessary) until the desired three digits of accuracy is achieved. In the present case, a third application of this procedure will be necessary and sufficient. Press [TRACE] and again move the trace cursor as close as possible to the high point on the curve. The values

$$X = 1.3744468 \quad Y = .15062644$$

displayed at the bottom of the screen are the approximate coordinates of the high point (x_0, y_0) on the curve. How do we know that x_0, and y_0 have at least three digits of accuracy? Choose points (x_1, y_1), (x_2, y_2), on the curve on both sides of the point (x_0, y_0). If x_1 and x_2 have three significant digits of agreement, if $x_1 < x_0 < x_2$, if $y_1 < y_0$, and $y_2 < y_0$, then x_0 is accurate to at least three digits.

It's possible for the value of y_0 to have less than three digits of accuracy. With the trace cursor back at (x_0, y_0) press [CLEAR] (to break the trace mode) ▲ ▲ to read

1.3. FUNCTIONS IN MATHEMATICS

the coordinates
$$X = 1.3744468 \quad Y = .15083375$$
of a point clearly above the high point. Press ▼ ▼ ▼ ▼ to read the coordinates
$$X\,1.3744468 \quad Y = .15041328$$
of a point clearly below the high point. Since the first three digits of the y-coordinates of these two points are unchanged, it follows that y_0 has three digits of accuracy.

To find the coordinates of the low point on the curve, we first return to the original graph so that we can see the approximate location of the low point. This means that we must reenter $X_{min} = -2\pi$ and $X_{max} = 2\pi$ in the Window Menu, and then choose ZoomFit from the Zoom Menu. A procedure similar to the above will produce
$$X = -1.40369 \quad Y = -.2688205$$
as the approximate coordinates of the low point on the curve with three assured digits of accuracy in the x, and y coordinates. ■

Before we leave this example, let us consider how rigorous our conclusions are. If we can trust our eye sight and the output of our calculator, it would appear that our conclusions are beyond question. Undoubtedly that is the case, but interestingly enough, there can be, as we shall see, **hidden twists and turns in graphs which are nearly impossible to find** unless we know they are there. (See **Problem 12** in the Exercise Set at the end of this chapter). Calculus will provide us with valuable tools to find these hidden features, and it will increase immeasurably the rigor of our work.

★★★★ 86 ★★★★

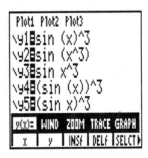

The big surprise is that the function $y1$ shown on the accompanying screen represents $\sin(x^3)$, and not the function $\sin^3(x)$ we wish to enter. Rather than start the argument for a function with a left parenthesis, the calculator uses an empty space. **Actually the parentheses enclosing x are redundant on this calculator.** Of the five functions appearing on this screen, **only $y4$, and $y5$ represent** $\sin^3(x) = (\sin(x))^3$. The functions $y2$, and $y3$, are equivalent to $y1$. We will discuss this in greater detail after this example.

While the solution to this example will be much the same on this instrument, the key strokes will be quite different, because the $y(x) =$ menu, along with the TRACE and ZOOM menus are all submenus of the GRAPH menu. The $y(x) =$ submenu has items for deleting functions and for activating and deactivating functions. When two menu bars are showing on the screen, remember to press the M-keys to use the top menu bar. After entering the expression and deleting or deactivating unrelated expressions from previous work, we access the WINDOW MENU by pressing [2ND] [M2], and the ZOOM MENU by pressing [2ND] [M3]. A ▶ symbol appearing after the last item on the lower menu bar indicates that this menu has additional items which are not displayed. Press [MORE] to access other items in the list. In the ZOOM MENU press [MORE] to find ZFIT, which can be used to size the graph for "best fit" after x_{min}, x_{max} have been set in the WINDOW MENU.

This device computes with 12 digits of accuracy—more than the TI-83. The output should agree with the above, although it will, of course, show more digits.

Finally, let us return to our discussion about parentheses that we started above. As we saw in the screen above, sin(x) can be entered in the form **sin x** without enclosing the x in parentheses. All of the special functions of mathematics stored on this device are formatted in the same way.

In the following table, tan(x) to used to represent a typical special function accessed from the keyboard. The top row contains various expressions as they would normally appear in printed form. Listed below each entry are various ways of entering the expression on the calculator.

$2\tan(x)$	$\tan(x)+2$	$\tan(2)+x$	$\tan(2+x)$	$\tan^2(x)$	$\tan(x^2)$
2*tan x	2+tan x	x+tan 2	tan (2+x)	(tan x)∧2	tan x∧2
tan x*2	tan x+2	tan 2+x	tan (x+2)	(tan (x))∧2	tan (x)∧2

In the first three columns, parentheses enclosing x or 2 are acceptable but unnecessary. As you can see, the first five columns show a consistent pattern which basically agrees with the meaning of parentheses in our pencil and paper world. The sixth column, however, is inconsistent with the first five, and it certainly runs counter to the usual meaning of parentheses.

This surprising problem seems to be quite isolated. It happens only when combining special functions accessed from the keyboard with the ∧ operator in this very specific way.

★★★★ 86 ★★★★

Example 1.2 *Solve the equation* $x^3 = 2 + 7\cos(x)$ *using a graphical technique. Solve it again using the calculator's equation solver.*

A solution to this equation is the x-coordinate of any point of intersection between the curves $y = x^3$ and $y = 2 + 7\cos(x)$, or alternately any x-intercept of the curve $y = x^3 - 2 - 7\cos(x)$. Electing the first approach, press [Y=], move the cursor to $Y1$, press [CLEAR] and enter $Y1 = X \wedge 3$. Enter the second function in the same way as $Y2$. Clear or deactivate any remaining functions in the Y= menu. The y-coordinate of the second function is clearly between -5 and 9. By looking at $y = x^3$ it follows that any intersection point would have to be between -2 and 3. Press [WINDOW] and enter $X_{min} = -2$, $X_{max} = 3$, $Y_{min} = -5$, $Y_{max} = 9$. Press [GRAPH] and we will see that there is exactly one point of intersection. Press [TRACE] and use ▶ and ◀ to move the trace cursor to the intersection. The ▲ and ▼ keys move the blinking point from one curve to the other. They can be used to help decide whether the point is to the left or right of the point of intersection. Use the ZOOM and TRACE menus as we did in the last example to locate the x-coordinate of the

1.3. FUNCTIONS IN MATHEMATICS

point of intersection. After zooming in twice, we find that $X = 1.4341755$ has three digits of accuracy. An approximate solution to the equation is $x = 1.43$.

A graphical strategy such as this should always include a statement concerning the degree of accuracy. Example 1.1 describes in detail how to gather enough evidence to make such a statement.

To use the calculator's equation solver, it is still a good idea to first graph the two curves (or the curve $y = x^3 - 2 - 7\cos(x)$) so that a good first guess can be given to the solver. **If an equation has multiple solutions, this is an essential step**, but it is always a good idea. A graph will also help to establish whether or not an equation has multiple roots.

After creating the initial graph as we did above, it is clear that there is a solution for some x between 1 and 2. Press [MATH] and scroll down to the tenth item and press [ENTER] (or just press [0]) to get the equation solver. To enter the equation, we use the **Y-VARS menu** to enter the formulas for $Y1$ and $Y2$ that we already entered in the Y= menu. Move the cursor to $eqn : 0 =$ and enter [VARS][1][1] to enter $Y1$ followed by [-][VARS][1][2] to subtract $Y2$. Enter $X = 2$ (or $X-1$) as a first guess for X and press [ALPHA]([SOLVE]), to get $X = 1.4344407234$. (Before [ALPHA]([SOLVE]) is pressed the cursor **must be at the location of the variable that the equation is being solved for.** ∎

★★★★ 86 ★★★★

This device has a more convenient solver. Press [2ND][SOLVER] to get the equation solver menu. A submenu appears with convenient access to all of the functions in the $y(x) =$ menu. Enter $y1 = y2$ at $eqn:$ by pressing [F1][ALPHA]([=])[F2][ENTER]. Enter a first guess of $x = 2$, and with the cursor at this location (the variable that the equation is being solved for), press the solve command [F5].

★★★★ 86 ★★★★

There are many details about the equation solver that have not been discussed with this example. **You are encouraged to read pages 2-8 through 2-12 in the TI-83 Guidebook** (★86★ pages 202-206 in the TI-86 Guidebook ★86★).

Example 1.3 *Graph the function defined piecewise by the formula*

$$f(x) = \begin{cases} 12 + 2x & x \leq -4 \\ 10 - x^2 & -4 < x < 4 \\ 3x - 6 & x \geq 4 \end{cases}$$

This function must be entered into the Y= menu in the following very precise way. Press [2ND][TEST] to gain access to the inequality symbols. Be careful to include all of the parentheses exactly as shown.

$\backslash Y1 = (12 + 2*X)(X \leq -4) + (10 - X \wedge 2)(x > -4)(x < 4) + (3*X - 6)(X > 4).$

The default mode for drawing graphs is **Connected**. Points ,are plotted, and the points are connected by straight line segments. If there are breaks on the curve, the calculator still insists on connecting the breaks with line segments, which produces an undesirable "look" to the graph. To avoid the problem press MODE (★86★ 2ND [MODE] ★86★) and change the mode to **Dot**. This, however, changes the mode permanently to Dot (until it is changed again). A more interesting approach changes the mode of just this particular function to Dot. Go into the Y= menu, and move the cursor to the symbol (usually a \ )just to the left of the symbol $Y1$. Press ENTER several times, scrolling through various modes, until a dotted backslash appears (★86★ in the $y(x)=$ menu, press MORE and then F3 —the style key—several times ★86★). ■

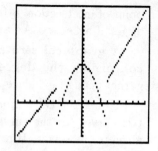

1.4 Applications

Example 1.4 *A parameter v controls the temperatures T_1 and T_2 of two probes inserted into an object according to*

$$T_1 = 14.2 + 6.1\cos(0.35v + 1), \quad T_2 = 9.8 - 2.4\sin(0.16v - 2),$$

where temperature is in centigrade degrees and $0 \le v \le 30$. Graph both functions. Determine (to three significant digits) the value v_0 of the parameter v which maximizes the difference in temperature from the first to the second probe. Create an "appropriate" table of values of T_1, and T_2 for a range of v values centered about $v = v_0$

Let x denote the parameter v, let $Y1 = T_1$, and $Y2 = T_2$. Press Y= , move the cursor to $Y1$, press CLEAR and enter the function T_1 . Repeat this step for $Y2$, and clear or deactivate any remaining functions on the calculator.

In order to see the maximum difference in temperature from the first to the second probe, we plot as well the function $Y3 = Y1-Y2$. The formulas for $Y1$ and $Y2$ that we already entered can be accessed through the Y-VARS menu. Move the cursor to $Y3$ and press VARS ▶ 1 1 to enter $Y1$. Press - VARS ▶ 1 2 to finish the entry for $Y3$. (★86★ Move the cursor to $y3$ and press F2 1 - F2 2 . ★86★) Set $x_{min} = 0$, and $x_{max} = 30$ in the WINDOW menu, choose ZFit from the ZOOM menu, and press GRAPH . Press TRACE and use ▲ ▼ to switch the blinking insertion point between the three curves. With the blinking insertion point on $Y3$ use ▶ and ◀ to move the point to the high point on the curve. Use the ZOOM and TRACE menus as we did in Example 1.1 to locate the x-coordinate of the high point on the curve to three significant digits of accuracy.

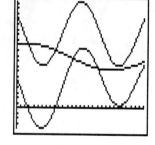

1.4. APPLICATIONS

To speed up the process, it will help to press [Y=] and **deactivate the functions Y1, and Y2**. The x-coordinate of the high point turns out to be 15.558511, and so we set $v_0 = 15.5$.

To set up the table of values press [Y=] switch $Y1$, and $Y2$ back to their active mode (so that they appear on our table), and deactivate $Y3$ (so that it does not appear on our table). Press [2ND] [TBLSET] and enter 15.5 (the value of v_0) for the value of **TblStart** (★86★ first press [TABLE] ★86★). As an interesting aside, we could also have entered the letter X. As long as we did not change the position of the trace cursor on the curve $Y3$, the value of this letter continues to be X=15.558511. In the present case, however, the three digit display is preferable.

Given the three digits of accuracy, an appropriate table of values would contain values for $Y1$ and $Y2$ for a range of x-values of the form $\ldots, 15.3, 15.4, 15.5, 15.6, \ldots$. Consequently, we set Δ Tbl=0.1. We are now ready to display our table. Press [2ND] [TABLE], and use ▲ and ▼ to scroll backwards and forwards to other parts of the table. Press ▲ ▲ ▲ to move the position of v_0 to the center of the table. ■

Example 1.5 *An oval running track (two semicircles connected by straight line segments as shown below) must be 440 yards long. Plans are made to turn the enclosed rectangular area into a playing field. What dimensions should the track have to maximize the area of the enclosed rectangle.*

Let L be the length of the straight line segment, and let r be the radius of the semicircle. Using, among other things, $2\pi r$ as the circumference of a circle, we have

$$length = 2\pi r + 2L = 440, \quad area = 2rL.$$

We let $X = r$ on our calculator. After solving the first equation for L, we let $Y1 = 220 - \pi x$ ($Y1 = L$) in the Y= menu. The area of the enclosed rectangle can then be entered as $Y2 = 2xY1$. Recall that $Y1$ can be entered by pressing [VARS] ▶ [1] [1].

We wish to find the high point on the graph of $Y2$, but first we need to establish a range of values for X. The formula for $Y2$ is well defined for all real numbers X, but clearly negative values for X make no sense in this applied situation. We could let $X = 0$ (the track could be a 220 yard line segment), and X could be large enough that $L = 0$ (the track could be a circle—the other extreme). In that case the radius X satisfies $2\pi X = 440$ or in other words, $X = 220/\pi$. Press [WINDOW], enter $X_{min} = 0$, $X_{max} = 220/\pi$, and then press [ZOOM] [0] (for ZoomFit). The graph will appear with its high point roughly in the center. After zooming in three times using the methods of the last three examples, we arrive at a high point at $X = 35.014087$, $Y = 7703.0992$, with three digits of accuracy, in both X and Y. The maximum area is $A = 7700 \, m^2$ (the unit's digit is not significant). The radius is $r = 35.0 \, m$. To find the corresponding value for L, leave the blinking point at the high point on the curve so that X continues to be $X = 35.014087$, and press $\boxed{\text{2ND}}$ $\boxed{\text{QUIT}}$ to go the the home screen. The value of $L = Y1$ is found by pressing $\boxed{\text{VARS}}$ $\blacktriangleright$ $\boxed{1}$ $\boxed{1}$ $\boxed{\text{ENTER}}$. This gives $Y1 = 110.0000000$, so that $L = 110 \, m$ to three places.

To indicate the number of digits of accuracy, answers are frequently formatted in scientific notation

$$L = 1.10 \times 10^2 \, m, \; r = 3.50 \times 10^1 \, m, \; A = 7.70 \times 10^3 \, m^2.$$

We could have arrived at this answer more analytically, and more precisely. The area function A is a quadratic. By looking at the shape of the field corresponding to $r = 0$, and $r = 220/\pi$ ($L = 0$) it is clear that $A = 0$ at both extremes. It follows that the vertex of the quadratic, in this case a high point, is midway between $r = 0$ and $r = 220/\pi$. Consequently, $r = 110/\pi = 35.01408748$. ■

1.5 Limits of Functions

The idea of a limit lies at the very foundation of calculus. Indeed, calculus could be regarded as the study of limits. Suppose that $f(x)$ is defined on some interval centered at $x = a$, except possibly at $x = a$ itself. Whether $f(a)$ is defined or not is of no importance to the idea of a limit. All that matters is that $f(x)$ is defined for all other points x sufficiently close to $x = a$. Intuitively, $\lim_{x \to a} f(x) = L$ means that $f(x)$ gets closer and closer to L as x gets closer and closer to $x = a$, always choosing $x \neq a$. It means that $f(x)$ can be made to be as close to L as we want, simply by choosing x close enough to a. A more rigorous definition will be discussed later. Our graphics calculator is an excellent tool to study both definitions.

Example 1.6 *Use graphical evidence to determine the probable value of* $\lim_{x \to 2} f(x)$ *for* $f(x) = \frac{x^2 + 7x - 3}{3x - 2}$

Actually, this example is quite straight forward, and solvable without the need of a calculator. Using the *Limit Theorem* involving limits of sums, products, and quotients from any main text, it is an easy matter to show that the limit is $15/4 = 3.75$. On an even more intuitive level, clearly, if $x = almost(2)$, then

$$\frac{x^2 + 7x - 3}{3x - 2} = \frac{almost(4) + almost(14) - 3}{almost(6) - 2} = almost(\frac{15}{4}).$$

Nevertheless, our calculator is a good tool to emphasize the idea of a limit. Press $\boxed{\text{Y=}}$, enter $Y1 = f(X)$ and clear or deactivate all of the other functions in the Y= menu. Press $\boxed{\text{WINDOW}}$ and choose values for X_{min} and X_{max} so that $X_{min} < 2 < X_{max}$. Press $\boxed{\text{ZOOM}}$ $\boxed{0}$ (for ZoomFit) and the graph will appear.

1.5. LIMITS OF FUNCTIONS

Press [TRACE] and move the trace cursor as close as possible to $X = 2$. Immediately, we begin to gather evidence concerning the limit. As X approaches 2, the values of $Y1$ can also be read off the bottom of the screen. On this first screen, X cannot get very close to 2, but already, the units digit for $Y1$ appears to have leveled off to a value of 3. With X as close as possible to 2, we force X to be closer yet to 2 by zooming in on the graph at this point. Press [TRACE] and repeat the analysis. Watch the values $Y1$ as X gets still closer to 2. As X gets close to 2, more of the digits for $Y1$ will level off and remain unchanged.

This experiment could be repeated infinitely many times (with enough patience), but before long we will have reached the limits of accuracy for our calculator. Well before that happens, $Y1$ will stabilize to a value of $Y1 = 3.75$ ■

Have we proved anything? **Certainly not!** We would have to zoom in infinitely many times in order to verify the value of a limit. Obviously, we can never do this, and so we will never be able to use graphical ideas to verify a limit. On the other hand, the evidence is overwhelming that the limit exists and has a value of $L = 3.75 = 15/4$.

Limits are much more interesting when they cannot be evaluated in such an easy way. If $f(x) \to 0$ and $g(x) \to 0$ as $x \to a$, then

$$\lim_{x \to a} \frac{f(x)}{g(x)} \text{ is said to have indeterminate form } \frac{0}{0}.$$

Such a limit cannot be evaluated directly. Letting x be close to a produces a quotient of two small numbers which can have any value depending on the relative smallness of the two numbers. A Theorem, *L'Hopital's Rule*, offered later in this course, will allow us to compute limits of this sort analytically, but for now, our calculator will be an indispensable tool to establish (probable) values for these limits. Interestingly enough, we will soon see that our calculator can also fail fundamentally on this kind of a problem.

Example 1.7 *Determine the (probable) value of* $\lim_{x \to 4} f(x)$ *for*

$$f(x) = \frac{\sqrt{x} - 2}{x - 4}.$$

by creating a table of values $(x, f(x))$ *where, the entry x gets as close as we want to 4 by going far enough out in the table.*

Notice that this limit is a (0/0) indeterminate, and that $f(4)$ is undefined, so we cannot determine the limit of this expression by observation.

We created a table earlier in the chapter. Enter $Y1 = f(x)$ into the Y= menu, and deactivating all other functions. There are several ways to set up meaningful tables, but the requirements in this example call for some special tricks.

First, we discuss a different (more straight forward) table. Press [2ND] [TBLSET] (⋆86⋆ press [TABLE] ⋆86⋆), and set TblStart= 4, and ΔTbl= 0.1. Press [2ND] [TABLE] to see the table. Use the ▲ and ▼ keys to see entries on both sides of $X = 4$. The entry for $X = 4$ will, of course, read ERROR, because $f(4)$ is undefined.

This table does not contain X-values which are very close to $X = 4$, so press $\boxed{\text{2ND}}$ $[\![\text{TBLSET}]\!]$ again, return TblStart= 4, set ΔTbl= 0.01, and press $\boxed{\text{2ND}}$ $[\![\text{TABLE}]\!]$. Repeat this process for a sequence of values ΔTbl= $0.01, 0.001, 0.0001, \ldots$.

The table called for in the example will allow us to see all of this in one table, but we need to adjust X so that we can let $X \to 4$ by scrolling up or down the table. We can't do that in the table we created above. One way to do this is to replace x by $x_n = 4 + \frac{1}{n}$. Then, $x_n \to 4$ from the right hand side of 4 as $n \to +\infty$, and $x_n \to 4$ from the left hand side of 4 as $n \to -\infty$. We could then create a table of values which shows $(x_n, f(x_n))$ for $n = \pm 1, \pm 2, \ldots$

To speed up the process, we could replace x instead by $x_n = 4 + \frac{1}{n^2}$, which goes to 4 faster. Unfortunately, this x_n is always larger than 4, so we would only be determining a limit of $f(x)$ from the right hand side of 4. Better still, we could let $x_n = 4 + \frac{1}{n^3}$, which goes to 4 faster yet, and goes there from the right or left depending on whether $n \to +\infty$ or $n \to -\infty$. This is the approach we use.

Of course, we only have the letter X to use on our calculator, so we will be replacing X by $4 + \frac{1}{X^3}$.

Return to the $Y=$ menu. Leave $Y1$ as it was entered above. Enter $Y2 = 4 + 1/X \wedge 3$, and $Y3 = Y1(Y2)$. Recall that $Y1$ and $Y2$ are entered from the Y-VARS menu (press $\boxed{\text{VARS}}$ $\blacktriangleright$ $\boxed{1}$ $\boxed{1}$ for $Y1$). ($\star 86\star$ Enter the letter y from the $y(x) =$ menu, and use 1 and 2 from the keyboard. $\star 86\star$) Then deactivate $Y1$, since we don't need to see it in our table.

Finally, go to the TBLSET menu and set TblStart= 1, and ΔTbl= 1. Press $\boxed{\text{2ND}}$ $[\![\text{TABLE}]\!]$ to see the table. Recall that if x is an entry in the $Y2$ column, then the corresponding entry in the $Y3$ column will be $f(x)$. Scroll up or down to see entries to the left or right of $x = 4$. **To speed up the process even more, set ΔTbl in the TBLSET menu to a integer larger than 1**, and the entries in the $Y2$ column will converge to $x = 4$ even more quickly. That will hardly be necessary in the present problem. By scrolling up or down far enough, the entries in the $Y3$ column level off at 0.25. Based on this table, it **appears** that

X	Y2	Y3
10	4.001	.24998
11	4.0008	.24999
12	4.0006	.24999
13	4.0005	.24999
14	4.0004	.24999
15	4.0003	.25
16	4.0002	.25

X=16

$$\lim_{x \to 4} \frac{\sqrt{x} - 2}{x - 4} = .25 = 1/4,$$

but again, it is important to realize that this work does not have the strength of a proof. ∎

This limit can also be determined by analytic means. See your main text for the appropriate algebraic manipulations involved.

Example 1.8 *Determine the probable value of the following limit by graphical means.*

$$\lim_{x \to 0} \frac{\cos(2x) - 1 + 2x^2}{x^4}.$$

1.5. LIMITS OF FUNCTIONS

Here is another example of a (0/0) indeterminate form. We will not be able to determine this limit by analytic means until much later in the course.

After entering the expression, start with a window determined by $X_{min} = -1$, $X_{max} = +1$ and ZoomFit. Pressing [TRACE] will cause a minor irritation. The function does not have a value at $X = 0$, and so the trace key will not move the trace cursor to the curve. Press [CLEAR] (to break the trace) and use ▲ and ▼ to move the point to the curve. Otherwise the example is much like Example 1.6. Zoom in once to get $Y = .66666062$ when $X \neq 0$ is close to 0. Zoom in again, and the graph shows signs of chaos. Zoom in one more time and the graph becomes dramatically chaotic.

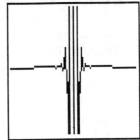

Something is clearly wrong here! It is not at all obvious, but the limit value $L = 1/12 = 0.08333333$ happens to exist. Our attempt to determine this, however, has certainly failed. The fault lies not with our calculator, but with unavoidable problems dealing with decimal approximations. When decimal arithmetic is performed on a fraction of the form A/B, where both A, and B are small, round off errors become significant. They are unavoidable! Calculators could be built with more digits of accuracy, but regardless of how many digits are used, all we have to do is arrange for A and B to be small enough, and decimal arithmetic on the value of A/B will fail to be accurate. We will not be able to determine the value of this limit until much later in the course. ■

In spite of this problem, the above methods frequently will determine the values of limits, including one-sided limits, limits where $x \to \pm\infty$, and limits where $f(x) \to \pm\infty$. Because of the ever present possibility of round off errors, however, we must accept that these methods can sometimes fail.

While the intuitive definition of a limit is very useful, it is not very precise. What do the words "close" and "close enough" mean? With this definition, we would be able to handle intuitive ideas, but we would not be able to take mathematics beyond the intuitive. The following more precise definition is needed, not only for this reason, but for philosophical reasons as well.

We say that $\lim_{x \to a} = L$ if given any $\epsilon > 0$ there is a $\delta > 0$ such that

$$|f(x) - L| < \epsilon \text{ whenever } 0 < |x - a| < \delta.$$

The term $|f(x) - L|$ is the distance between $f(x)$ and L, and $|x - a|$ is the distance between x and a. Our calculator can be used effectively to gain an understanding of this difficult definition.

Example 1.9 Let $f(x) = \frac{\sin(6\pi x)}{\sin(2\pi x)}$. Determine the probable value of $L = \lim_{x \to 1} f(x)$. Let $\epsilon = 0.0001$, and determine the corresponding number δ.

This is another example of a 0/0 indeterminate form. Enter $Y1 = f(x)$ into the Y= menu, and use the graphical method of Example 1.6 or the table method of Example 1.7 to establish that $L = 3$.

If we find a δ, then any smaller δ will work as well. To make this search more interesting, we should try to find the best δ. What is most interesting about δ, however, is usually not it's exact display of digits, but rather its order of magnitude. That is, we are usually more interested in deciding which of the numbers $\delta = 0.01, 0.001, 0.0001, \ldots$ will satisfy the ϵ-condition.

To determine δ we graph $y = f(x)$ along with the horizontal lines $y = L \pm \epsilon = 3 \pm .0001$. Any point on the graph of $y = f(x)$ which lies between these two horizontal lines satisfies $|f(x) - L| < \epsilon$. Enter $Y2 = 3 + .0001$ and $Y3 = 3 - .0001$. In the WINDOW menu, set $Y_{min} = 3 - .0002$ and $Y_{max} = 3 + .0002$. Values for X_{min} and X_{max} will have to be adjusted so that they are close enough to $X = 0$ that an appropriate graph appears. Such a graph would show a portion of the graph which lies between the two horizontal lines, and a portion on both sides that lies outside the lines. After some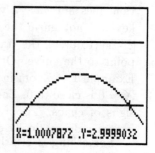
experimentation we choose $X_{min} = 0.999$, $X_{max} = 1.001$, and press [GRAPH] to get the screen shown. The values of x satisfying the ϵ-inequality lie between the two intersection points on the graph. We let δ be the distance from the limit point $x = 1$ to the end points of this interval. The limit point $x = 1$ may be centered in this interval but it doesn't have to be. In this case we choose the smaller distance for δ. Move the trace cursor to the intersection point on the right to get $X = 1.0007872$. We let $d_1 = X - 1 = .0007872$. Move the trace cursor to the intersection point on the left. It may be easier to compute $1 - X$ from the home screen. Press [1] - [X,T,θ,n] [ENTER] to get $d_2 = 1 - X = 7.782340426$E-4 or roughly .0007782. This display of digits is not of great concern, so we set $\delta = .0007$. ∎

1.6 Exercise Set

Many of the problems from your main calculus text serve as good practice problems, especially in this early chapter when one of the primary objectives is to get acquainted with the calculator. As an added advantage, answers are frequently supplied in the back of your main calculus text.

A graph is complete if all of its twists and turns, and its long term behavior outside of picture are all displayed in the picture. (Actually, we should say, "A picture of a graph is complete.")

1. Evaluate the following expressions as decimals.

   a) $3 \times \frac{6 \times 7^2 - 8}{5} - 2 + \frac{14}{6 \times 5 + 8}$ b) $\frac{8}{5} \times 3 - \frac{8}{5 \times 3}$

   c) $(-2^4 + \frac{1}{5^2})^{-1}$ d) $\frac{\frac{5}{3}}{14} - \frac{6}{\frac{2}{3}}$

   The answers rounded off to three decimal places:
   a) 169.968 b) 4.267 c) -0.063 d) -8.881

2. Express the answer to each part of Problem 1 in the form of a rational number a/b where a and b are integers. (Press [MATH] and look at the ▶Frac menu) (★86★ Press [2ND] [[MATH]] [F5] [MORE] [F1] to get the ▶ Frac menu. ★86★)

1.6. EXERCISE SET

ans: a) $\frac{16147}{95}$ b) $\frac{64}{15}$ c) $-\frac{25}{399}$ d) $-\frac{373}{42}$

3. Evaluate the following expressions.
    a) $\frac{\sqrt{43.7}}{5^2}$ b) $\frac{\sqrt[3]{-8}}{\sqrt{25}} + 81^{1/4}$
    c) $\frac{(73)^{2/3}+17}{(-3)^{1/5}}$ d) $\sin^2(-3\pi/7)$

4. Evaluate $8\sqrt{\cos(13°)}$ in the following ways:
    a) in radian mode, by mathematically transforming 13° to radian measure.
    b) directly, by pressing [2ND] [ANGLE] and using the ° menu item (★86★ first press [2ND] [MATH] ★86★).
    c) directly, by pressing [MODE] and converting the calculator to a degree mode of operation (★86★ just press [2ND] [MODE] ★86★).

5. Graph the function $f(x) = \frac{x^2+50x+442}{x^2+2}$. Find the x-intercepts, and the coordinates of the local maximums and minimums with at least three digits of accuracy. Are you convinced that you have a "complete" graph What are the y-values tending to as $x \to -\infty$? Verify your claim by creating a table of values for $x = -100, -200, \ldots$

6. Graph the function $f(x) = x^4 - 50x^3 - 28x^2 - 280x - 642$. Provide evidence that the graph is "complete". Find, with at least three digits of accuracy, the x-intercepts, and the coordinates of the local maximums and minimums.

7. Graph the function $f(x) = \begin{cases} x + 10 & x \leq -6 \\ 4 + 2\sin(\pi x) & -6 < x < 6 \\ 10 - x & x \geq 6 \end{cases}$

8. Graph the ellipse $\frac{x^2}{64} + \frac{y^2}{9} = 1$ by using $Y1$ and $Y2$ to represent its upper and lower branches.

9. Let $g(x) = \sin(a + x)$. Define $g_2(x) = g(g(x)), g_3(x) = g(g_2(x)), g_4(x) = g(g_3(x))$. In other words, define the compositions $g_2 = g \circ g, g_3 = g \circ g \circ g, g_4 = g \circ g \circ g \circ g$. Graph all four functions on the interval $0 \leq x \leq 4\pi$ for
    a) $a = 0$ b) $a = 1$ c) $a = \pi/2$ c) $a = 2$ e) $a = 3$ f) $a = \pi$
Strive for an easy input. If $Y1 = g(x)$, you can define $g_2(x)$ by $Y2 = Y1(Y1)$. Recall that the functions $Y1, Y2, \ldots$ can be accessed by pressing [VARS] ▶ [1] (★86★ y1, y2 ... can be accessed from the y(x) =menu by pressing [F2] followed by a number ★86★). Enter the letter A ($a = A$) directly into the formulation of $Y1$. Give A a particular value by entering, for example, [0] [STO▶] [ALPHA] ([A]) [ENTER] from the home screen.

10. The formula $C = \frac{5}{9}(F-32)$ gives Centigrade temperature in terms of Fahrenheit. Suppose that the average day time high temperature at a certain location is $T = 13 - 12\sin(\frac{2\pi}{365.25}(t + 61))$, where t is time in days since the first of the year, and T is in Centigrade degrees. Create a table showing the average day time high in both Centigrade and Fahrenheit degrees for the days corresponding to $t = 30, 60, 90, \ldots$

11. Find the largest and smallest value of $f(x) = x^4 + 2x^3 - 76x^2 - 242x + 100$ on the closed interval $[-9, 9]$.

12. **A graph with a hidden twist** Graph the function $f(x) = x^2 + (x+8)^{2/3}$ on the interval $-20 \leq x \leq 20$. Are you confident that you have a complete graph?

It turns out that there is a **dramatic sharp corner** at the point on the curve corresponding to $x = -8$. To see it, you will have to zoom in on the point many times (perhaps as many as 10). Rather than using the ZoomIn feature, create a small window more quickly by adjusting X_{min} and X_{max} in the WINDOW menu and using ZoomFit. After you have made the window small enough to see the corner, try to appreciate how small this window is relative to the original graph. Calculus will give us the means to graph functions more rigorously than we can with our eye sight.

13. A retail store has 2,000 feet of fencing to enclose three sides of a rectangular lot adjacent to a wall of the store. The store wall is 900 feet long, and no fencing is needed of this forth side. What dimensions maximize the enclosed area of the rectangular lot?

14. A computer company can expect to sell $q = 1200 - 0.32p$ computers per day if the selling price of a computer is p in dollars ($1000 \leq p \leq 3500$. What should the selling price be in order to maximize total revenue from sales? How many computers can it expect to sell at that price? What will the revenue be from sales? (Let $Y1 = q$, and $Y2 =$ revenue. This will help with the evaluations at the end of the problem. Recall how we can enter $Y1$ into the formula for $Y2$ by pressing $\boxed{\text{VARS}} \blacktriangleright \boxed{1} \boxed{1}$).

15. A rectangular pool-deck complex is being built for a hotel. The pool must have a water surface area of 4,000 square feet, and the pool must be enclosed on three sides by a 10 foot wide deck, and on the fourth side (the side facing the building) by a 30 foot wide patio. Find the dimensions which minimize the total area of the pool-deck complex.

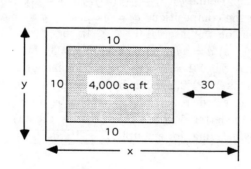

16. Determine the probable value of $\lim_{x \to 1} f(x)$ for $f(x) = \frac{\sqrt{x^2+x+14}-4}{x-1}$ using a graphical approach.

17. Determine the probable value of $\lim_{x \to 0} f(x)$ for $f(x) = \frac{\sin^2(6x)}{\sin(3x^2)}$ by creating a table of values $(x, f(x))$ where $x = 1/n^2$, $n = 1, 2, 3, \ldots$.

18. Compute, with four digits of accuracy, the following limits, if they exist in a finite or infinite sense.

1.6. EXERCISE SET

a) $\lim_{x\to 4}\left(\frac{3x^2-17x+20}{5x^2-19x-4}\right)$ b) $\lim_{x\to 3+}(2x^2\cot(\pi x))$ c) $\lim_{x\to 4}\left(\frac{2-\sqrt{x}}{x-4}\right)$
d) $\lim_{x\to 0}\left(\frac{5x-\sin(5x)}{x^3}\right)$ e) $\lim_{x\to\infty}\left(\frac{7x^2+3x-9}{2x^2+8x+5}\right)$

19. For the limit in Problem 16, for $\epsilon = .001$, and for $\epsilon = .00001$, find for each ϵ a $\delta > 0$ which will satisfy the ϵ-inequality in the formal definition of a limit.

Project: Swimming Coach

Suppose that you are the head coach for a large team of athletes who all plan to compete in a combination rowing/running event. The event begins on an island 2 miles out from a straight shore line. The end of the race is 10 miles down the shoreline, as shown in the picture below. Each competitor in the race will row to some *"transfer point"* along the beach, and then run from that point to the finish line. The choice of a transfer point is entirely up to each athlete. Simply by choosing the appropriate transfer point an athlete may decide to run the whole 10 miles or decide to completely avoid running, and any transfer point between (and including) these extremes is allowed.

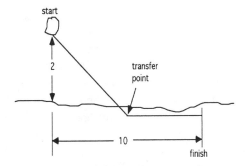

As the head coach, it is your responsibility to chose the best possible transfer point for each member of your team. Your only concern is that each athlete turn in his/her best total time. (Typical head coach attitude, right?) You know the rowing and running speeds of each member of your team, but these speeds change from day to day, and so what you need is a means of determining, quickly, and just before the race begins, a transfer point for each member of your team. You can take your calculator to the race as long as you careful not to drop it in the water. Let W, and R denote the rowing speed, and running speed in mph for any one of your athletes. Think of R and W as given constants for a particular athlete. Determine the the total time of the athlete's race, as a function of his/her transfer point X. Given values for R and W, this function can be graphed, and the best transfer point would be the x-coordinate of the low point on this curve.

The letters R and W can be used, from the keyboard, to enter into this total time function on your calculator. Before you can graph this function and find a low point for an athlete, you will need to give values to R and W. To give R, for example, a value of 9, press (from the home screen) $\boxed{9}$ $\boxed{\text{STO}\blacktriangleright}$ $\boxed{\text{ENTER}}$.

Nathan can row at 5.1 mph and run at 10.9 miles per hour. Amber can row at 6.2 mph and run at 9.7 mph. Determine their transfer points and their race times. (This is why they hired a mathematician as head coach.) Good luck!

It may help to split up the problem somewhat. Let $Y1$ be the swimming time, let $Y2$ be the running time. then $Y3 = Y1 + Y2$ is the total time.

Project: Road Construction

Towns A and B are separated by a river running basically east-west, and plans are being made to connect the towns by a road and a bridge over the river.

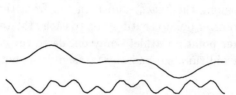

To locate everything precisely, a north-south and an east-west axis are placed over the region with units in kilometers (km). The origin is located at some point in the river as can be seen in the picture below. According to these axes, Town A is located 15 km west and 10 km north of the origin, and Town B is located 17 km east and 6 km south of the origin. The bridge can be built at any point along the river, but it must cross the river due north-south. To save money it is decided to build the road straight from Town A to the point P on the north side of the bridge, and straight again from the point Q on the south side of the bridge to Town B. (To simplify matters, assume that the road/bridge complex intersects the river bank only at P and Q.)

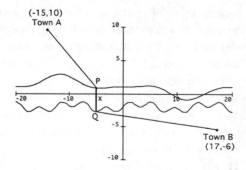

Cost, however, is a major complication. The north side of the river is hilly, expensive farm land, and it will cost $500,000 per km to build the road. The bridge itself will cost $2,000,000 per km to build. The south side of the river is

1.6. EXERCISE SET

flat, undeveloped, government land, and this part of the road can be built for only $190,000 per km. By trial and error it is determined that the curves representing the northern and southern boundaries of the river can be described reasonably well by the equations

$$y = 1 + 2\cos(x/4)\sin(x/7)^3, \text{ and } y = -2 + \cos(x/2)\sin(x),$$

where the northern boundary is listed first.

At what point along the river should the bridge be built in order to minimize the cost of the whole project? (It will help to split the problem up into several parts entered as $Y1, Y2, \ldots$).

Chapter 2

Differentiation

Calculus is the study of limits, and differentiation is one of the major limiting processes of calculus. Much of our effort on this topic will be spent mastering the skills of differentiating symbolically, and our calculator will not be very useful in this effort. Being able to differentiate, however, is of little value without a thorough understanding of the underlying concepts, and our graphics calculator can be used very effectively to study these important concepts. Our calculator can also compute numerical values for derivatives with relative ease, so it can be used wherever values for derivatives are needed.

2.1 The Definition

After zooming in at a point on the graph of a function several times, we have noticed that graphs of functions typically begin to look almost like nonvertical straight lines. This does not always happen, but it does happen whenever the function is differentiable at the point we are zooming in on. In fact, this property is the heart and soul, the beginning and the end, of what we mean by a differentiable function.

While differentiable functions can be quite complicated, they are, on the other hand, not so complicated. Look inside a small enough window, and they look like nonvertical straight lines. What could be simpler! As long as we restrict our analysis to small windows, the full weight of our understanding about straight lines can be applied to differentiable functions. This is why differentiable functions are important to mathematics, and this linear connection can lead to powerful insights.

We start our study from this graphical point of view. Let us take a function $f(x)$, zoom in on a point $(x_0, f(x_0))$ until it looks linear, and use that window to create the straight line. If this line happens to be nonvertical, then its slope is essentially $f'(x_0)$, the value of the derivative at x_0. Vertical lines have no slope—this is why they are not included in our discussion. Also, vertical lines are not functions of the underlying variable x—another good reason for excluding them from our discussion.

Example 2.1 *Let $f(x) = x\sin(x) + 2\cos(x)$. Find a window containing the point $(2, f(2))$ inside of which the graph of $f(x)$ "looks" linear. Find the slope m of the "line". Write the equation of the line through $(2, f(2))$ with slope m. Graph $f(x)$*

together with this line on the interval $0 \leq x \leq 10$ under ZoomFit. Zoom in on the point $(2, f(2))$ until the graph of $f(x)$ and the line coalesces into one.

Enter the function with $X_{min} = 0$ and $X_{max} = 10$ and choose ZoomFit. Press [TRACE], move the trace cursor as close as possible to $X = 2$ and Press [ZOOM] [2] [ENTER]. It will take three applications of this to see the linear appearance, and one more to make it a "sure thing.". **Press the trace key again**. It is important to get an actual point on the curve, and not one that looks like it's on the curve. It doesn't matter which point it is. As we have mentioned before, we can now switch to the home screen, and the values of the letters X and $Y1$ will represent the coordinates of the trace cursor on the graph. The point of tangency is $(2, f(2))$. It can be entered on the home screen in a similar way. To get $f(2)$ just enter $Y1(2)$ using the Y-VARS menu as we usually do to get $Y1$.

Enter the expression (Y1-Y1(2))/(X-2) on the home screen and press [STO▸] [ALPHA] [M] [ENTER]. The slope of the line is the value of M. The line through (x_0, y_0) having slope m is $y = y_0 + m(x - x_0)$. Consequently we can enter the sought after line in the Y= menu as $Y2 = Y1(2) + M*(X-2)$. Return to $X_{min} = 0$ and $X_{max} = 10$, choose ZoomFit, and see the original graph with its tangent line at the point $(2, f(2))$ on the curve.

It is an interesting exercise to now zoom in again on the point $(2, f(2))$. After zooming in three or four times, the original curve and its tangent line will become indistinguishable. ■

In this example, notice what we did mathematically to get the slope $M \approx f'(2)$ of the line. This process certainly complements the formal definition, that a function $f(x)$ is differentiable at a point a if

$$f'(a) = \lim_{x \to a} \frac{f(x) - f(a)}{x - a}.$$

or equivalently

$$= \lim_{h \to 0} \frac{f(a + h) - f(a)}{h}$$

exists as a finite number. The value $f'(a)$ is the slope of the line tangent to the graph at the point $(a, f(a))$ on the graph.

Example 2.2 *Compute $f'(\pi/4)$ for $f(x) = \tan(x)$ by setting up a table of approximate slope values which tends to $f'(\pi/4)$ in a manner similar to Example 1.7.*

Enter $f(x) = \tan(x)$ into the Y= menu as $Y1$. Move the cursor to the (=) sign and press [ENTER] to deactivate this function (we don't want to see its values) (⋆86⋆ with the cursor somewhere in y1 press [F5] from the $y(x) =$ menu ⋆86⋆). Then enter $(Y1 - Y1(\pi/4))/(X - \pi/4)$ as $Y2$. Since $f(x) = \tan(x)$ is so easy to enter, we could have skipped the first step and simply entered $(\tan(x) - \tan(\pi/4))/(x - \pi/4)$ directly as our first entry in the Y= menu. If the function $f(x)$ required more

2.1. THE DEFINITION

keyboard activity to enter, however, it would be much more convenient to enter these two expressions separately as we have done.

We want $X \to \pi/4$ from the right as we move down the table, and we want $X \to \pi/4$ from the left as we move up the table. To do this, we replace X by $X = \pi/4 + 1/t^3$, which goes to $\pi/4$ from the left and right quite quickly as t moves through the numbers $t = \pm 1, \pm 2, \ldots$. Since we only have the letter X to work with, we replace X by $\pi/4 + 1/X^3$ by using $Y3 = \pi/4 + 1/X \wedge 3$ to generate the new X-values. Finally, enter $Y4 = Y2(Y3)$. Deactivate $Y2$, Set TblStart = 1, ΔTbl=1 in the TBLSET menu, and press [2ND] [TABLE]. Recall that if x is an entry in the $Y3$-column, then $\frac{f(x)-f(\pi/4)}{x-\pi/4}$ is the corresponding entry in the $Y4$-column. **Scroll up or down** the table to see that as the $Y3$-values tend to $x = \pi/4 = .785398$, the $Y4$-values tend to a value of $f'(\pi/4) = 2$. ∎

We show next an example of a function $f(x)$ which is not differentiable at a point $(x_0, f(x_0))$. Such a function would have no nonvertical tangent line at $(x_0, f(x_0))$, or in other words, it would never look almost like a nonvertical line in small windows containing the point $(x_0, f(x_0))$. In our example, the point $(x_0, f(x_0))$ happens to be a sharp corner, but we could just as well have created an example with a vertical tangent line at $(x_0, f(x_0))$

Example 2.3 *Show graphically that $f(x) = x^2 + (x-2)^{2/3}$ is (probably) not differentiable at $x = 2$ by zooming in at $(2, 4)$ ($f(2) = 4$) often enough to make the reasonable claim that its graph will probably always have a sharp corner at this point regardless of how often we zoom in.*

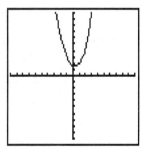

We must remember that the graphical evidence we collect will never carry the force of an analytical proof. Words like "probably" should always accompany a graphical argument like this.

After entering this function, start with a standard viewing screen by pressing [ZOOM] [6]. As you can see in the first screen, the graph looks quite typical. Zoom in once at $(2, 4)$ and the curved graph looks like its moving towards a straight line. Zoom in again, however, and the hint of a sharp corner at $(2, 4)$ begins to appear. The second screen shown is the result of zooming in four times. The sharp corner is displayed even more dramatically as the zooming-in process continues. ∎

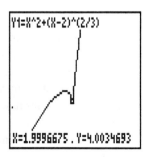

The techniques we have been using in this chapter to compute (approximate) numerical values for derivatives are important techniques to experience. Using them can give us a deeper understanding of what a derivative is both graphically and numerically.

2.2 Computing Derivatives

Our calculator can compute an approximate numerical value for a derivative in a more direct fashion. Press [MATH] [8] from the home screen or the Y= screen (among others) and the command $nDeriv(\ )$ will appear on that screen (★86★ press [2ND] [[CALC]] [F2] to enter the command $nDer(\ )$ ★86★). If $g(U)$ represents an expression in the letter U, and u_0 is a real number, then the input statement $nDeriv(g(U), U, u_0)$ **on the home screen will compute an approximate numerical value of $g'(u_0)$** when the ENTER key is pressed. An **optional** forth argument ϵ can be inserted to specify the **degree of accuracy**. The definition of this command is

$$nDeriv(g(U), U, u_0, \epsilon) = \frac{g(u_0 + \epsilon) - g(u_0 - \epsilon)}{2\epsilon}.$$

Smaller values of ϵ naturally produce more accuracy. If this argument is not inserted into the command, the default value of $\epsilon = 10^{-3}$ is used. Most of the time we will be using the letter X rather than U or any other letter, but the command was introduced in this way to indicate that any letter can be used.

Notice that $nDeriv(g(U), U, u_0)$ will produce a value even if the function $g(U)$ is not differentiable at $U = u_0$. Let $Y1$ be the function $f(x) = x^2 + (x - 2)^{2/3}$ in Example 2.3, and $nDeriv(Y1, X, 2) = 4$, even though we showed in that example that $f'(2)$ did not exist. Before this command can be used in a meaningful way, it must be established, independently, that $g'(u_0)$ exists. For the function $f(x)$ in Example 2.3, it turns out that $nDeriv(f(X), X, 2) = 4$ even though $f'(2)$ does not exist.

Example 2.4 *Let $f(x) = 3x^5 - 8x^2 + 9x + 6$. Compute the numerical value of $f'(5)$ using the calculator's command and the default value of ϵ. Compute it again using $\epsilon = 10^{-5}$. Compute $f'(x)$ symbolically, and evaluate $f'(5)$ exactly.*

Enter $f(x)$ as $Y1$ and, for the sake of comparison, enter its symbolic derivative $f'(x) = 15x^4 - 16x + 9$ as $Y2$. From the home screen, enter $nDeriv(Y1, X, 5)$ by pressing [MATH] [8] [VARS] [▶] [1] [1] [,] [X,T,θ,n] [,] [5] [ )]. Press [ENTER] to get $f'(5) \approx 9304.00075$. Enter $nDeriv(Y1, X, 5, 10 \wedge -5)$ in a similar way, and press [ENTER] to get $f'(5) \approx 9304.000000$. The exact value of $f'(5)$ can be obtained by evaluating $Y2(5)$. Since this is a rational expression (involving not too many digits), an exact value can be determined. Press [VARS] [▶] [1] [2] [( ] [5] [ )] [ENTER] to get $f'(5) = 9304$. ∎

★★★★ 86 ★★★★

We have mentioned on several occasions that values such as $y2(5)$ can be entered from the $y(x) =$ menu, or directly from the keyboard in a natural way. There is no [VARS] key on this device. The CALC menu contains not only the $nDer(\ )$ command for computing the approximate value of a derivative, but it has as well a $der1(\ )$ and a $der2(\ )$ command for computing the actual first and second derivatives of a function at a point. This allows us to get exact numerical values for rational expressions (involving not too many digits). It is also interesting to point out that

2.2. COMPUTING DERIVATIVES

for the function $f(x)$ in Example 2.3, which is not differentiable at $x = 2$, we get $nDer(f(x), x, 2) = 4$ (as we have said, this command always evaluates), but entering $der1(f(x), x, 2)$ generates an ERROR message.
★★★★ 86 ★★★★

Example 2.5 Let $f(x) = x^7 + 9x^4 - 7x^2 + 8x + 2$ and let $g(x) = \frac{8x^3+16}{x^4+4}$, and let $h(x) = f(x)g(x)$. Demonstrate the Product Rule for derivatives at $x = -3$ by computing $h'(3)$ directly, and again by using the Product Rule applied to $f(x)$ and $g(x)$.

The Product Rule states that $h'(x) = f'(x)g(x) + f(x)g'(x)$. Press $\boxed{Y=}$, enter $f(x)$ as $Y1$, $g(x)$ as $Y2$, and enter $Y1 * Y2$ as $Y3$. On the home screen, enter $nDeriv(Y1, X, -3) * Y2(-3) + nDeriv(Y2, X, -3) * Y1(-3)$ and press $\boxed{\text{ENTER}}$ to get a value of -9145.706296. Enter $nDeriv(Y3, X, -3)$ to get a value of -9145.703896. Using $\epsilon = 10^{-5}$ as an optional forth argument in the derivative commands would have given us a higher level of agreement. (★86★ **Using the $der1$ command instead of $nDer$ should produce exact agreement.** ★86★) ∎

Example 2.6 Let $f(x) = \sqrt{x^2 + 6}$, let $g(x) = x^5 + 2x^3 + 8$, and let $h(x) = f(g(x))$. Demonstrate the Chain Rule for derivatives at $x = 4$ by computing $h'(4)$ directly, and again by using the Chain Rule applied to $f(x)$ and $g(x)$. Arrange for a higher level of agreement between the two answers by using $\epsilon = 10^{-5}$.

The Chain Rule implies that $h'(4) = f'(g(4))g'(4)$. Enter $f(x)$ as $Y1$, $g(x)$ as $Y2$ and set $Y3 = Y1(Y2)$. Recall that the $nDeriv$ command is in the MATH menu (★86★ the CALC menu ★86★). We compute $h'(4)$ directly by entering (on the home screen) $nDeriv(Y3, X, 4, 10 \wedge -5)$, and pressing $\boxed{\text{ENTER}}$ to get 1375.99693. To use the Chain Rule, enter $nDeriv(Y1, X, Y2(4), 10 \wedge -5) * nDeriv(Y2, X, 4, 10 \wedge -4)$ to get 1376. ∎

A function $f(x)$ is said to be increasing on an interval if $f(x_1) \leq f(x_2)$ whenever $x1$, x_2 are in the interval and $x_1 \leq x_2$.

Example 2.7 Graph the function $f(x) = x^2 + 20\sin(x)$ and its derivative $f'(x)$ using the calculator's numerical derivative command. Use a different graph style for each function in order to identify each function. Compare and discuss the significance of the pair of graphs.

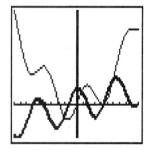

Enter $f(x)$ as $Y1$ in the $Y=$ menu, and then enter $nDeriv(Y1, X, X)$ as $Y2$. The X appearing in the second argument indicates that $Y1$ is to be differentiated with respect to the letter X. The X appearing in the third argument specifies that the derivative is to be evaluated at X which represents the location of the plotting cursor moving between X_{min} and X_{max}. It is this X in the third argument which makes $Y2$ behave like a function representing $f'(x)$. Move the cursor to the style \ symbol appearing just to the left of the symbol $Y2$, and press $\boxed{\text{ENTER}}$ repeatedly until the style changes to a bold \ symbol. This will generate a bold curve style for the derivative. (★86★ Look for STYLE by pressing [MORE] from the $y(x) =$ menu.★86★) The graph shown is the result of setting $X_{min} = -10$, $X_{max} = 10$, and choosing ZoomFit.

This pair of graphs contains a wealth of geometric information. Notice how the graph of $f'(x)$ is below the x-axis ($f'(x) < 0$) in the intervals where $f(x)$ is a decreasing function, and $f'(x)$ is above the x-axis ($f'(x) > 0$) in the intervals where $f(x)$ is an increasing function. Additionally, the graph of $f'(x)$ crosses the x-axis ($f'(x) = 0$) exactly at the points where the graph of $f(x)$ has a local high point or low point. ∎

The closing symbol marks an end to this example, but we could go on. We saw in Example 2.3 and even more dramatically in Problem 12 on page 15 that graphs of functions can have hidden twists and turns which are surprisingly small and impossible to see unless their existence is known in advance. Could the above graph contain a hidden twist or turn which is two small to see? If it did, the graph of $f(x)$ would have to switch between an increasing and decreasing graph in some small interval. Regardless of how small that interval is, it would have a dramatic effect on $f'(x)$ which would have to suddenly switch between positive and negative values. We are likely to see such a dramatic change in values on the above graph.

We are likely to see such a dramatic change in values of $f'(x)$, but it could still escape our attention. For example, if $f'(x)$ switched from positive to negative values and then back to positive values over an interval $a \leq x \leq b$ which lies between two adjacent pixels, then the negative values would not show on the screen.

The best way to see such hidden twists and turns in $f(x)$ is to compute $f'(x)$ symbolically. A formula for $f'(x)$ will always indicate where such a dramatic change in values occurs if there is such a point.

Example 2.8 *Let $f(x) = \frac{4x^2 - 20}{x^2 + 4}$ Determine all points on the graph of $f(x)$ where the tangent line passes through the point $P(8, 6)$.*

Enter $f(x)$ as $Y1$ in the $Y=$ menu, and show the graph of $f(x)$ in a window large enough to contain the point P. A standard window (press [ZOOM] [6]) will work reasonably well.

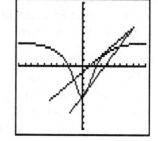

Move the cursor on the graph screen as close as possible to the point $P(8,6)$. Are there any points on the graph of $f(x)$ where the tangent line passes through P? There appears to be two such points, and we can **use the DRAW menu to actually draw (approximations for) the two lines.** Press [2ND] [DRAW] [2] ("Line(") (★86★ press [GRAPH] [MORE] [F2] [F2] ★86★). Press [ENTER] and the calculator will move back to the graph screen. With the cursor as close as possible to P, press [ENTER]. Press and hold ◄ (a line begins to form), and then press and hold ▼ until the line appears to be tangent to one of the two points on the graph. Then press [GRAPH] to fix the line on the graph. There is a second point of tangency on the graph and it can be drawn in the same way. **These lines can be erased by choosing ClrDraw in the DRAW menu.** From this graph, we can see approximately where these points of tangency are, and we will use the SOLVE menu to actually find them.

Compute $f'(x)$ symbolically, and enter $f'(x) = \frac{72x}{(x^2+4)^2}$ as $Y2$ in the Y= menu. The equation of the line through $(8,6)$ with slope m is $y = 6 + m(x-8)$. Let (a, b) denote one of the two points of tangency, so that $b = f(a)$. If (a, b) is a point on this line, then $b = f(a) = 6 + m(a-8)$.. If the line is also tangent to the graph at (a, b) then $m = f'(a)$. This gives the equation $f(a) = 6 + f'(a)(a-8)$, which can be solved for a using the SOLVE menu.

Press [MATH] and choose the tenth item (Solver ...). The equation must be entered in the form $0 = 6 + f'(a)(a-8) - f(a)$. It would be easier to enter the equation from the keyboard if we used X instead of A ($A = a$), but to avoid confusion, we will use the letter A to represent the unknown. In the solver menu, look for **eqn:0=** (it may be necessary to press ▲) and enter $6 + Y2(A) * (A-8) - Y1(A)$. Press [ENTER] or ▼ and bypass $X =$ (a default entry which is of no concern to us). Press ▼ to move the cursor to the line $A =$ and enter a start up value for A.

The letter A represents the x-coordinate of one of the points of tangency. From the graph we can see (with the help of the graph cursor) that one of these points has an x-coordinate of approximately 3.4. We use this as a start up value. **With the cursor on the line containing the unknown letter A that we are solving for**, press [ALPHA] ([SOLVE]) to see A change from its start up value to $A = 3.532423841808$. Return to the graph to see that the second point of tangency has an x-coordinate of approximately .4. Enter this number as a start up value for A and use the same process to get $A = .32888124011307$. ■

2.3 Exercise Set

1. Show graphically that the function $f(x) = \frac{x^2-9}{x^2+9}$ appears to be differentiable at $x = 4$, by zooming in on the point $(4, f(4))$ until the graph "looks" linear. Find an approximate value for $f'(4)$ by finding the slope m of this line. Write the equation of the line through the point $(4, f(4))$ having slope m, and enter the resulting linear function into the Y= menu. Graph both $f(x)$ and this line on a larger window. Zoom in again on the point $(4, f(4))$. What happens to the two graphs as you zoom in? Compare the value of $m \approx f'(4)$ to the approximate value of $f'(4)$ determined using the $nDeriv$ command. Compare you answer by differentiating symbolically.

2. Create a Table of values, using the definition of a derivative as the limit of a certain difference quotient, to compute $f'(4)$ for $f(x) = \sqrt{x^2 + 9}$. Use one of the methods suggested in the examples, or or a method of your own to get $x \to 4$ (or $h \to 0$) in your table. Compare your answer with the answer determined by using the $nDeriv$ command. Compare your answer once more by differentiating symbolically.

3. Compute the derivatives of the following functions at the specified point by using the definition of a derivative and by using the $nDeriv$ command.

a) $f(x) = x^2\sqrt{4-x^2}$, $x = 3/2$

b) $g(w) = \sqrt{1+\sqrt{1+\sqrt{1+w^2}}}$, $w = 5$

c) $H(t) = \sec^5(\frac{t^2}{t^2+9})$, $t = 2$

4. Verify that the *Product Rule* holds for the derivative of $h(x) = f(x)g(x)$ at $x = 7$ if $f(x) = x^4 + 5x^3 - 8x^2 + 12$ and $g(x) = \sqrt{x^3 + 14x^2 + 8}$. Compute all of the derivatives directly from the definition of a derivative. Check your answers using the *nDeriv* command.

5. Verify that the *Quotient Rule* holds for the derivative of $h(x) = \frac{f(x)}{g(x)}$ at $x = -3$ if $f(x) = \tan(5x+9)$ and $g(x) = x^2 + 10x = 7$. Compute all of the derivatives directly from the definition of a derivative. Check your answers using the *nDeriv* command.

6. Verify that the *Chain Rule* holds for $h(x) = f(g(x))$ at $x = 7$, if $f(x) = \frac{3x+1}{\sqrt{x^2+9}}$ and $g(x) = x\cos(2x)$.

7. Let $f(x) = 3 + 4x + |2x - 7|$. Graph $f(x)$ and find the point $x = a$ where $f'(a)$ does not exist. Zoom in on the point often enough to be convinced that the "corner will never round out". Compute $nDeriv(f(X), X, a)$. Look at the definition of this command on page 24 to explain the reason for its value. By creating a table of values, observe why the definition of a derivative as the limit of a certain difference quotient fails to produce a limit at this point. Use one of the methods suggested in the examples, or a method of your own to get $x \to a$ (or $h \to 0$) in your table.

8. Let $f(x) = 3 + x + \sqrt[3]{4x^4 - 20x^3 - 143x^2 + 420x + 1764}$. Graph $f(x)$, and find all points x where $f'(x)$ does not exist. By zooming in on each of the points, describe in specific geometric terms why the derivative fails to exist. Compute $nDeriv(f(X), X, a)$ at each of these points $x = a$. At each point, by creating a table of values, observe why the definition of a derivative as the limit of a certain difference quotient fails to produce a limit.

9. Let $f(x) = \frac{4-\sqrt[5]{x^4-6x^2+9}}{x^2+5x-8}$. Graph $f(x)$, and find all points x where $f'(x)$ does not exist. There should be four such points and two of them may be hard to find. A symbolic analysis of parts of the derivative $f'(x)$ will help to find them. By zooming in on each of the points, describe in specific geometric terms why the derivative fails to exist. Compute $nDeriv(f(X), X, a)$ at each of these points $x = a$.

10. The proof that $\frac{d}{dx} = \cos(x)$ for an arbitrary value of x follows by showing that

$$(i) \lim_{h \to 0} \frac{\sin(h)}{h} = 1, \text{ and } (ii) \lim_{h \to 0} \frac{1 - \cos(h)}{h} = 0.$$

Establish these limit values numerically by creating a table of values (one table for both) in which $h \to 0^+$ as you continue going down in the table, and $h \to 0^-$ as you continue going up in the table.

11. Find the points on the graph of $y = x^2 + 23\sqrt[3]{x}$ whose tangent line passes through the point (5,13). (Use the line command in the DRAW menu to estimate the location of the points.)

2.3. EXERCISE SET

12. Steps are being taken to reintroduce the rare, purple bellied flitter bird into the wild. Wild life biologists predict (somewhat hopefully) that its population will follow the curve
$$P(t) = \frac{19t^2 + 40 - 48t \sin(t/2)}{t^2 + 200},$$
where t is time in years after the introduction, and $P(t)$ is population in thousands. How many birds were initially released into the wild? At what rate is the population expected to change after 4 years, after 10 years, after 20 years? Reductions in the population are expected as the bird competes in the wild. At what rate is the population expected to decrease midway through its first period of decline? What is the final population when it stabilizes. These questions can be answered in varying degrees of rigor. Describe the methods used to arrive at your answers.

Chapter 3

Applications of Differentiation

The applications of differentiation are virtually endless. In this chapter, we discuss a range of applications to mathematics itself, and to the applied scientific world. Given the strong connection between a derivative of a function and the geometry of its graph, it should not be surprising that our first application is to graphing. However, given our ability to graph functions by pressing a few keys on a calculator, you might think that our graphing skills are in need of no further refinement. Such is not the case.

3.1 Graphing Polynomials

Polynomials are among the most straight forward of all mathematical functions, and one would expect their graphs to be quite simple. Surprisingly enough, for polynomials beyond degree 2 and 3 this is not the not generally the case. A polynomial $p(x)$ always goes to $\pm\infty$ as $x \to \pm\infty$ and $p(x)$ usually gets large rather quickly. The first term of the polynomial $p(x) = x^5 - x^2$, for example, has a value of 100,000, when $x = 10$, overwhelming the rest of the polynomial.

If we just graph a polynomial on the interval $-10 \leq x \leq 10$, (and whose to say that this interval is big enough) the graph will quickly leave the plotting window, if we choose ZStandard in the ZOOM menu. If we choose ZFit, the scale will be so large that the graph will be completely flat except for its tails which would get positively or negatively large. Under ZoomFit and fairly normal values for X_{min} and X_{max}, all polynomial graphs "look like" one of the four simple graphs $f(x) = ax^n$ with $a = \pm 1$ and $n = 2$ or 3.

Example 3.1 *Graph the polynomial*

$$p(x) = x^5 + 325x^4 + 4022x^3 - 16965x^2 - 45500 * x + 8500.$$

Supply evidence that you have the complete graph (that there are no hidden turns, which are not displayed).

Enter $p(x)$ as $Y1$, set $X_{min} = -10$, $X_{max} = 10$, and choose ZoomFit. The resulting graph is quite a disappointment! What are we to do? One approach is to press WINDOW and reduce the size of $Ymin$ and Y_{max} until a reasonable shape appears. Setting Y_{min} and Y_{max} equal to ± 200000, helps somewhat and produces the screen shown.

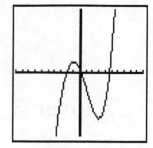

Do we have a complete graph? **With this large of a scale, the values of $p(x)$ could rise and/or fall several hundred points and we would not be able to see it on this screen.** The best way to resolve these critical issues is to study the sign of $p'(x)$. We know that $p(x)$ increases on intervals where $p'(x)$ is positive, and $p(x)$ decreases on intervals where $p'(x)$ is negative. This analysis can be done on several levels. We could plot $p'(x)$ by entering $nDeriv(Y1, X, X)$ as $Y2$. This might help a great deal, but $p'(x)$ is another polynomial, and its graph could have its own set of problems. A more rigorous approach is to compute $p'(x)$ symbolically. We enter

$$p'(x) = 5x^4 + 4 \cdot 325x^3 + 3 \cdot 4022x^2 - 2 \cdot 16965x - 45500$$

as $Y2$. The **first attempt to graph** $p'(x)$ shows two zeros at approximately $x = -1$, and $x = 3$, and $p'(x)$ looks **decidedly nonzero** everywhere else. This could well be a complete graph of $p'(x)$, but under ZoomFit, $p'(x)$ appears to be headed for another zero to the left of $x = -10$. Consequently, we plot $p'(x)$ under $X_{min} = -15$, $X_{max} = 5$, and ZoomFit. As you can see in the screen shown, a third zero of $p'(x)$ occurs near $x = -12$.

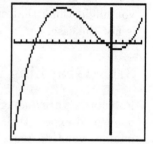

Look carefully at this graph, and you can see that **something is missing**. A fourth degree polynomial with a leading term of the form ax^4 ($a > 0$) would have to tend to $+\infty$ as $x \to \pm\infty$. Consequently, $p'(x)$ must turn around once more, somewhere to the left of $x = -15$, and this would produce a fourth zero.

It is difficult to find this zero graphically, but we can compute this hidden zero and all of the other zeros of $p'(x)$ with considerable accuracy by using the SOLVER menu and the approximate values of the zeros showing in the graph. Press MATH 0 to get the solver menu. Enter $Y2$ at $eqn : 0 =$, press ▼, and enter one of the three approximate zeros $x = -12, -1, 3$ for $X =$. With the cursor remaining on the line '$X =$, press ALPHA (SOLVE). Repeat this procedure for the next two zeros to get

$$x_1 = -11.778733214564, \; x_2 = -1.0147139206323, \; x_3 = 3.04214734170884$$

How do we find the remaining zero? We know it is smaller than -15, so we blindly enter numbers less than -15 on the line $X =$, until we get a answer different from the above three. Enter a start-up value of $X = 200$ on this line and the forth zero of $p'(x)$ is determined to be $x_4 = -250.24902628187$.

By looking at the positive and negative values of $p'(x)$, we conclude that the graph of the original function $p(x)$ increases on the intervals

$$-\infty \leq x \leq x_4, \; x_1 \leq x \leq x_2, \; x_3 \leq x \leq +\infty,$$

and $p(x)$ decreases on the interval
$$x_4 \leq x \leq x_1, \; x_2 \leq x \leq x_3.$$

The original graph shown above is now seen to be a complete graph of $p(x)$ at least on the interval $-10 \leq x \leq 10$. Because of the large numbers involved, it is hard to get a useful picture which would include all four critical points of $p(x)$. ∎

3.2 Graphing Functions

When we graph functions on a graphing calculator, we can only show the graph on a finite and usually small window. One way to talk precisely about the long term behavior of a graph outside of a finite window is to talk about the asymptotic behavior of a graph. The line $y = L$ is said to be a horizontal asymptote of $f(x)$ if $f(x) \to L$ as $x \to -\infty$ or as $x \to +\infty$. The line $y = mx + b$ is said to be a slant asymptote if
$$\lim_{x \to -\infty} (f(x) - (ax+b)) = 0, \text{ or } \lim_{x \to +\infty} (f(x) - (ax+b)) = 0.$$

Additionally, if $f(x) \to \pm\infty$ as $x \to \pm\infty$, we can use the graphs of the well understood power functions $p(x) = ax^n$ to help describe the behavior of many other functions outside of a finite window.

We say that $f(x)$ **has order of growth** $p(x) = ax^n$ if
$$\lim_{x \to -\infty} \frac{f(x)}{p(x)} = 1, \text{ or } \lim_{x \to +\infty} \frac{f(x)}{p(x)} = 1.$$

If we let $u(x) = f(x)/p(x)$, then $f(x) = u(x)p(x)$ and $u(x) \approx 1$ when x is large. This is the sense in which $f(x)$ "looks like" $p(x)$ when x is large.

The order of growth of a polynomial is the leading term of the polynomial. To see this, let $f(x) = a_0 + a_1 x + \ldots + a_n x^n$, and $p(x) = a_n x^n$, and write
$$\frac{f(x)}{p(x)} = \frac{a_0}{a_n x^n} + \ldots + \frac{a_{n-1}}{a_n x} + 1.$$

It follows that $\lim_{x \to \pm\infty} (f(x)/p(x)) = 1$.

In the same way, if $f(x)$ is the rational function
$$f(x) = \frac{A(x)}{B(x)} = \frac{a_0 + a_1 x + \ldots + a_n x^n}{b_0 + b_1 x + \ldots + b_m x^m}$$
with $n > m$, then $f(x) \to \pm\infty$ as $x \to \pm\infty$, and $f(x)$ has order of growth
$$p(x) = \frac{a_n x^n}{b_m x^m} = \frac{a_n}{b_m} x^{n-m}.$$

When $n = m$, the line $y = L = a_n/b_m$ is a horizontal asymptote, and when $n = m+1$, the rational function has a slant asymptote which can be found by long division. In this case, the quotient $Q(x)$ will be linear,
$$f(x) = Q(x) + \frac{R(x)}{B(x)}, \text{ and } \lim_{x \to \pm\infty} \frac{R(x)}{B(x)} = 0,$$
because the degree of the remainder $R(x)$ is smaller then m, the degree of $B(x)$.

Example 3.2 Graph the function $f(x) = \frac{x^3-25x}{x^2+4}$. Find the intervals were $f(x)$ is increasing, decreasing, concave up, and concave down. Find the coordinates of the local maximums and minimums. Determine the asymptotic behavior of $f(x)$.

Long division leads to $f(x) = x - \frac{29x}{x^2+4}$, so $y = x$ is a slant asymptote. We enter $f(x)$ as $Y1$ and its derivative

$$f'(x) = \frac{x^4 + 37x^2 - 100}{(x^2 + 4)^2}$$

as $Y2$, and the asymptote $p(x) = x$ as $Y3$. We deactivate $Y2$ and graph $Y1$ along with its asymptote just to demonstrate how the graph of $f(x)$ approaches its asymptote. In a ZStandard zoom window, the local maximum and minimum points of $f(x)$ are seen to occur near $x = \pm 2$.

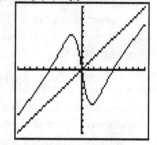

There are no other local maximum or minimum points. At such a point we would have $f'(x) = 0$, which would imply that its numerator $N(x) = x^4 + 37x^2 - 100 = 0$. Notice that $N(x)$ clearly increases as $|x|$ increases (all of the terms involving x are squared and positive). If $N(x_1) = 0$, then $N(x) > 0$ for all x with $|x| > |x_1|$. Furthermore, If $N(x_1) = 0$, then $N(-x_1) = 0$.

Using the SOLVE menu, as we did in the last example, to find one solution to $f'(x) = 0$, we get both solutions $x = \pm 1.5905159052374$. The local maximum and minimum are $(\pm 1.5905159052374, \mp 5.473311335)$.

The sign of $f'(x)$ must be constant in each of the three intervals determined by these two points, because $f'(x)$ is continuous and has no other zeros. Consequently, the above two x-values separate the real line into three intervals in which $f(x)$ definitely increases or decreases as shown in the graph.

To determine the intervals of concavity, we must investigate the sign of $f''(x)$ in much the same way as we did $f'(x)$. A formula for the second derivative would be tedious to compute, so we will take advantage of the *nDeriv* command to analyze this expression. Move the cursor to $Y4$ in the $Y=$ menu and press $\boxed{\text{MATH}}$ $\boxed{0}$ to enter $Y4 = nDeriv(Y2, X, X)$. The graph of $f''(x)$ appears to by quite small except for x near 0, and the only clear zero of $f''(x)$ is at $x = 0$. It appears to be negative to the left of 0 and positive to the right of 0.

Is $f(x)$ concave down for all $x < 0$, and concave up for all $x > 0$? Look at the graph of $f(x)$ again and notice that there appears to be a subtle shift to concave up in the far left hand side of the window, and a subtle shift to concave down in the far right hand of the window. **More interesting, remember that $y = x$ is an asymptote, and so there would have to be such a change in concavity in order for the graph of $f(x)$ to approach the asymptote.** Change the values of Y_{min} and Y_{max} in the WINDOW menu to ± 2 and graph $f''(x)$ again to see that there is indeed a subtle change in the sign of $f''(x)$ at roughly $x = \pm 3.4$.

Surprisingly enough, we can solve the equation $Y4 = 0$ using the SOLVE menu. It is **hard to say how accurate this would be**, since $Y3 = nDeriv(Y2, X, X)$ is

3.3. THE MEAN VALUE THEOREM

itself a numerical approximation with only three decimals of accuracy. Nevertheless, the SOLVE menu produces the zeros $x = \pm 3.464$ of $f''(x)$, and the obvious zero at $x = 0$. It follows that $f(x)$ is concave down on the intervals [-3.464,0] and [3.464,∞], and concave up on the intervals $[-\infty, -3.464]$ and [0,3,464]. (★86★ Using $der1$ instead of $nDer$ for $Y4$ will produce more accurate roots of $f''(x) = 0$. Both of these commands are available in the CALC menu. ★86★) ■

The **probability of making mistakes is quite large** whenever we enter functions that involve such excessive keyboard activity. Certainly, we should check the input screens for signs of mistakes, but even with such diligence, mistakes are bound to escape our attention from time to time. **To use technology successfully, we must maintain a skeptical attitude. Does the output look correct and reasonable?** Frequently it is easy to see if a mistake has been made. For example, if $f(x)$ and $f'(x)$ are both entered as functions in the Y= menu, compare their graphs on the same screen. The values of $f'(x)$, and the shape of the graph of $f(x)$ are connected in such as visual way, that a make a mistake in one of the entries should be immediately observable.

3.3 The Mean Value Theorem

Example 3.3 *Let $f(x) = 5 + e^x + 12\sin(x)$ for x in the interval $[a, b] = [-3, 4]$. Show graphically that the Mean Value Theorem $f(b) - f(a) = f'(c)(b - a)$ is satisfied at three points c in the interval $[a, b]$. Find the points using the SOLVE menu.*

Graphically, the *Mean Value Theorem* says that for at least one c in the interval $-3 < c < 4$, the tangent line to the graph at $(c, f(c))$ is parallel to the secant line through the points $(-3, f(-3))$ and $(4, f(4))$.

Enter $f(x)$ as $Y1$, and $f'(x) = e^x + 12\cos(x)$ as $Y2$. Deactivate $Y2$, set $X_{min} = -3$, $X_{max} = 4$, and view the graph under ZoomFit. It would help to adjust the vertical size of the window by making Y_{min} somewhat smaller. A value of $Y_{min} = -20$ was used to produce the screen shown. The bottom of the screen is frequently cluttered with X and Y coordinate values, and this maneuver keeps the bottom of the curve out of this mess.

To draw the secant line, it helps to access the DRAW menu from the graph screen. After pressing [2ND] [DRAW] [2], the graph screen returns. Move the cursor to one of the two points $(-3, f(-3))$, $(4, f(4))$ and press [ENTER]. Move the cursor to the other point and press [ENTER] again. To finalize the resulting line, press [GRAPH]. By looking at the screen shown, it is clear that there are three parallel tangent lines at points c in the interval. Press [TRACE] and move the trace cursor to the approximate locations to get the candidates $x = -0.9, 1.4, 2.9$.

To find the slope of the secant line, we enter $(Y1(4) - Y1(-3))/7$ on the home screen using the Y-VARS menu. After it is entered, press [STO▶] [ALPHA] [M] to give M the value $M = 6.737167588$. (★86★ Pressing [STO▶] automatically activates the **alpha mode**. After the expression is entered, simply press [STO▶] [M]. ★86★)

In the SOLVE menu, enter $eqn : 0 = Y2 - M$ and solve the equation using the above start-up values that we read off of the graph. This gives us the solutions $c = -1.0108908874751, 1.3180156087818, 2.9138310142436$. ∎

Example 3.4 *Let $f(x) = 8 - x^2 + 26(x - 2)^{2/3}$ for x in the interval [-1,5]. Show graphically that $f(x)$ does not satisfy the Mean Value Theorem on this interval, and discuss which conditions of the hypothesis of the theorem are satisfied by $f(x)$ and which are not.*

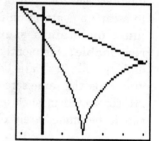

Enter $f(x)$ as $Y1$, $Z_{min} = -1.1$, $X_{max} = 5.1$, and graph $f(x)$ under ZoomFit. Draw the secant line between $(-1, f(-1))$ and $(5, f(5))$ as we did in the last example using the DRAW menu. From the screen shown, it is clear that no tangent line will be parallel to the secant line. ∎

Notice that $f(x)$ is continuous on the interval [-1,5] and differentiable everywhere in the interval except at $x = 2$ where a sharp corner is located. As you can see, even a slight weakening of the hypothesis of the *Mean Value Theorem* can lead to a result which is no longer true.

3.4 Newton's Method

Finding an exact solution to an equation is frequently an impossible task. While mathematicians speak confidently about the existence of solutions, actually finding exact solutions is another matter. Frequently, solutions cannot be found, even though they clearly exist, and this happens to be the destiny of mathematics, and not a reflection of the person doing the work.

Surprisingly, even the solution to an equation as straightforward as a polynomial may be impossible to find. A formula for solving a second degree polynomial equation is well known to every student of algebra, and formulas exist, as well, for solving third and fourth degree polynomial equations, although they are less well known. The Norwegian mathematician Niels Abel proved in 1824, however, that there is no formula for solving a general fifth degree polynomial equation, and later the French mathematician Evariste Galois proved that there is no formula for solving a general polynomial equation of degree n for any $n \geq 5$. By a formula here, we mean a formula based on root taking and the operations of $+, -, \times, \div$ which would be applied to the coefficients of the polynomial. Perhaps even more striking, it can be shown, for example, that not even one root of the equation

$$x^5 - 9x + 3 = 0$$

can be expressed by root taking and applying the operations of $+, -, \times, \div$ to integers (or rational numbers).

Thus, in order to solve most equations, we must rely instead on approximation techniques. As we are all aware, our graphing calculator has an excellent command

3.4. NEWTON'S METHOD

for solving equations, so why do we need any further study on this topic? There are several answers to this question. The most important answer is that we must understand how and why mathematics works before we can use it effectively. We must remain in control of technology rather than the other way around. Additionally, it is interesting to point out that the **equation solving menu on our calculator doesn't always work**. This may well be a rare event, but if it fails, there are other techniques that might work. This interesting situation is encountered in our first example.

One of the most well known of these techniques is *Newton's Method*. In order to solve the equation $f(x) = 0$, we make an initial guess $x = x_0$, determine the tangent line to the graph of $y = f(x)$ at the point $(x_0, f(x_0))$ on the graph, and then find the point $x = x_1$ where this line crosses the x-axis. It is a straight forward process to find this line, set $y = 0$, and solve for x to get

$$x_1 = x_0 - \frac{f(x_0)}{f'(x_0)}.$$

Usually, the point $x = x_1$ is closer to a solution of $f(x) = 0$ than the initial guess $x = x_0$. If we use $x = x_1$ as the new initial guess, the same process leads to yet a better estimate $x = x_2$. Repeating this process over and over again leads to the iterative formula

$$x_{n+1} = x_n - \frac{f(x_n)}{f'(x_n)},$$

known as *Newton's Method*. If $x_n \to a$ as $n \to \infty$, then $x = a$ **always** solves the equation $f(x) = 0$. To see this, notice that if $x_n \approx x_{n+1}$, then

$$x_n \approx x_{n+1} = x_n - \frac{f(x_n)}{f'(x_n)}$$

means that $f(x_n)/f'(x_n) \approx 0$, and hence $f(x_n) \approx 0$. In practice, we calculate x_n until x_n and x_{n+1} agree to the desired number of digits, and then we use $x = x_n \approx x_{n+1}$ as the approximate solution to the equation.

While *Newton's Method* is usually reliable, it can fail spectacularly. We will show how this can happen in a moment, but first let us try an example where it works.

Example 3.5 *Use Newton's Method to find the intersection points of the curves*

$$y = f(x) = 25x^4 - 214x^2 - 120, \quad y = g(x) = -60x^3 + 600x + 240.$$

Compare the solutions obtained by Newton's method to those obtained using the SOLVE menu. Notice that the **SOLVE** *menu fails to find one of the solutions that is successfully determined using Newton's Method.*

Newton's Method can only be applied to an equation in the form $h(x) = 0$, so define $h(x) = f(x) - g(x) = 25x^4 + 60x^3 - 214x^2 - 600x - 360$. Enter $h(x)$ as $Y1$ and $h'(x) = 100x^3 + 180x^2 - 428x - 600$ as $Y2$. Finally, enter

$$Y3 = X - Y1/Y2.$$

In order to get start up values for the solution to $h(x)=0$, we look at the graph of $h(x)$. First deactivate $Y2$ and $Y3$, set $X_{min} = -10$, $X_{max} = 10$ (a blind guess), and graph $h(x)$ under ZoomFit. A disappointingly flat graph will appear—we could have anticipated this. Remember, however, that we don't really need to see a complete graph, but only one that shows all of the points where $h(x)$ crosses the x-axis. With this in mind, press [WINDOW] and reduce the sizes on Y_{min} and Y_{max}, until we see these points. With $Y_{min} = -1000$, $Y_{max} = 1000$ we can see roots near $X = \pm 3$, and one **or possibly two** near $x = -1$. The graph appears to be large outside of this window, but we should be careful about making such an assessment.

On the home screen, press [3] [STO▸] [X,T,θ,n] [ENTER] to give X the start up value of $x_0 = 3$. The value of x_1 in the *Newton's Method* iterative procedure is $Y3$. We give X the value of $Y3$ by entering [VARS] [▸] [1] [3] [STO▸] [X,T,θ,n] [ENTER], on the home screen. With this preliminary work, we can now easily continue. **Press [2nd] [[ENTRY]] to re-enter the last input statement**, and press [ENTER] again to get a value for x_2. Press the combination [2ND] [[ENTRY]] [ENTER], repeatedly to get successive values $x_3, x_4, \ldots$, until the output numbers are no longer changing. This happens quickly, with the terms leveling off at $x_4 = 3.16227766$. Let us compare this answer to the answer we get using the SOLVE menu. Press [MATH] [0] and enter the equation $Y1 = 0$ and $X = 3$ as a start up value. Press [ALPHA] [(SOLVE)] to get $X = 3.16227766\ldots$.

This process can be repeated with $x_0 = -3$ to get an approximation of $x_4 = -3.16227766$ from *Newton's Method* which will agree with the answer obtained using the SOLVE menu.

Using the start up value of $x_0 = -1$, however, presents an interesting difference in performance in the two approaches. Using $X = -1$ in the SOLVE menu produces a solution of $X = 3.16227766$—**clearly not the desired solution!** Does this mean that there is no solution to $h(x) = 0$ near $x = -1$? **We can zoom in on this point on the graph of $h(x)$. It will always appear to touch, but this question can never be resolved by zooming in.** No matter how much we zoom in, perhaps $h(x)$ will separate from the x-axis if we zoomed in still more. Using *Newton's Method* with a start up value of $x_0 = -1$, however, leads eventually to a solution of $x_{17} = -1.1999996565$. If there is any doubt, evaluate $Y1$ at this value of X by pressing [VARS] [▸] [1] [1] to get $Y1 = 0$. It should be said, however, that if *Newton's Method* converges, it always converges to a solution to the equation. ∎

This doesn't make *Newton's Method* better than the SOLVE menu on our calculator. *Newton's Method* can also fail, and on the whole, the SOLVE menu on our calculator is probably more often reliable.

Finally, **a keyboard shortcut** should be mentioned. On the home screen, if nothing is entered on the active line, then pressing just the [ENTER] key, is equivalent to pressing [2ND] [[ENTRY]] [ENTER]. Instead of pressing the combination [2ND] [[ENTRY]] [ENTER] repeatedly, it is enough to just press [ENTER] repeatedly.

Example 3.6 *Let $f(x) = \frac{13x^2 - 25}{2x^2 + 10}$. Show graphically that $f(x) = 0$ has solutions near $x = \pm 1$. Show that Newton's Method fails (spectacularly) if $x_0 = 5$ is used as a start up value. Show that it fails again, with $x_n \to \pm\infty$ if $x_0 = 5.001$ (or if x_0 is any*

3.5. APPLICATIONS

other start up value with $|x_0| > 5$). Show that the method converges to a solution if $x_0 = 4.999$ (or any other number with $|x_0| < 5$. Draw a picture to explain why it fails at $x_0 = 5$.

Enter $f(x)$ as $Y1$ and $f'(x) = \frac{40x}{(x^2+5)^2}$ as $Y2$. Enter $Y3 = X - Y1/Y2$ to generate the terms of the sequence in *Newton's Method*. Deactivate $Y2$ and $Y3$, set $X_{min} = -7$, $X_{max} = 7$, and graph $f(x)$ under ZoomFit. The two zeros near $x = \pm 1$ are clearly visible in the graph.

On the home screen, press $\boxed{5}$ $\boxed{\text{STO}\blacktriangleright}$ $\boxed{\text{X,T,}\theta\text{,n}}$ $\boxed{\text{ENTER}}$ to give X a value of $x_0 = 5$. As we discussed in the last example, press $\boxed{\text{VARS}}$ $\blacktriangleright$ $\boxed{1}$ $\boxed{3}$ $\boxed{\text{STO}\blacktriangleright}$ $\boxed{\text{X,T,}\theta\text{,n}}$ $\boxed{\text{ENTER}}$ to get a value of the next term $x_1 = -5$ in Newton's sequence. Press $\boxed{\text{2ND}}$ $\boxed{\text{ENTRY}}$ $\boxed{\text{ENTER}}$ to get the second term $x_2 = 5$. It may be entertaining to continue, but there is no need to do so. Since $x_2 = x_0$, the terms will now alternate between ± 5 forever.

To see why this happens, let us return to the geometry of *Newton's Method*. The values $x_0 = 5$ and $x_1 = -5$ imply that the tangent line to the graph at $(5, f(5))$ crosses the x-axis at $x = -5$. The values $x_1 = -5$ and $x_2 = 5$ imply that the tangent line to the graph at $(-5, f(-5))$ crosses the x-axis at $x = 5$. We draw these two tangent lines. We can open up the DRAW menu from the home screen or from the graph screen. On the graph screen, press $\boxed{\text{TRACE}}$ and move the trace cursor as close as possible to $(5, f(5))$. Press $\boxed{\text{2ND}}$ $\boxed{\text{DRAW}}$ $\boxed{5}$ $\boxed{\text{ENTER}}$ to see the line appear on the graph. Enter the tangent line at $(-5, f(-5))$ in the same way to get the screen shown.

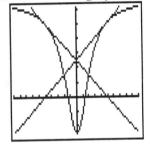

Look at this graph and the geometry which underlies *Newton's Method* and you can see why $x_n \to \pm \infty$ if a start up value x_0 with $|x_0| > 5$ is used. If we give X the value $x_0 = 5.001$ and compute $x_1, x_2, \ldots$ as we did before, then the first few values of x_n stay close to ± 5, and then they rapidly get large with $x_7 = -1119.002136$. If we choose $x_0 > 5$ to be even closer to 5, then more terms of the sequence will stay close to ± 5, but as soon as x_n moves somewhat away from ± 5, the terms get large rapidly.

Finally, give X the start up value of $x_0 = 4.999$, and the terms level off at $x_{10} = -1.386750491$. Notice that a positive start up value determines a negative solution. In the same way, $x_0 = -4.999$, yields a solution of $x_{10} = 1.386750491$. ∎

3.5 Applications

In applied mathematics, we are frequently interested in finding the largest or smallest value of some function. Examples of this sort were considered in Chapter 1, where we used strictly visual criteria to locate the point of interest. These same problems can be solved more rigorously with the powerful tools of calculus.

Example 3.7 *An Air Traffic Controller is monitoring the positions of two airplanes on a radar screen. Both are flying at the same altitude. At a certain moment, the passenger jet is 75.4 miles South of the airport and traveling North at 547 miles per hour. At the same moment, the air force fighter jet is 126 miles Northeast of the* airport and traveling Southwest at 714 miles per hour. Neither plane will be landing at the airport, and both will be flying over the airport with no change in speed or direction. How close do the planes get to each other? What is the moment of closest approach, and where are the two planes at this moment. Create a table of values in 10 second intervals centered at the moment of closest approach, of the rate of change in the distance between the planes in miles per hour.

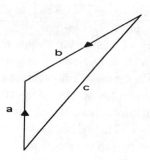

The *Law of Cosines* gives us the equation $c^2 = a^2 + b^2 - 2ab\cos(\alpha)$. The angle α is 135°. We let t be time in hours with $t = 0$ corresponding to that "initial moment," and we wish to minimize c as a function of t.

$$a = a(t) = 75.4 - 547t, \quad b = b(t) = 126 - 714t, \quad c = c(t) = \sqrt{a^2 + b^2 - ab\cos(135°)}).$$

Naturally, we must use X to represent the variable t. We enter a as $Y1$, b as $Y2$. The angle that the jet fighter is approaching the airport is given in degrees. We could use the MODE menu to switch our calculator over to degree measure, but it is just as easy to leave it in radian measure and transform the angle of 135° into $3\pi/4$. As a convenience we enter $\cos(3\pi/4)$ on the home screen and press [STO▶] [ALPHA] ([s]) [ENTER] to give S the value $S = \cos(3\pi/4)$. We could determine c directly as a function of t, but **it requires much less keyboard activity** if we enter c in the form

$$Y3 = \sqrt{(Y1 \wedge 2 + Y2 \wedge 2 - 2 * Y1 * Y2 * S)}.$$

Clearly, there is also a **much smaller risk of making a keyboard mistake** if c is entered in this way.

As the planes pass over the airport the values of a and b will turn from positive to negative values. This situation needs to be considered because the minimum distance between the planes could occur after one of the planes passes the airport. At first glance it is not clear that the *Law of Cosines* stated above is still valid for negative values of a and/or b. However, a moment's thought will convince you that as exactly one of the variables becomes negative, and then as they both become negative, the angle α changes from 135° to 45° and then back to 135°, introducing in the process just the right combination of negative signs to make the formula valid for all values of a and b. The key idea here is that $\cos(135°) = -\cos(45°)$

3.5. APPLICATIONS

We could deactivate $Y1$ and $Y2$ and look only for the low point on the graph of $Y3$, but it might be more interesting if we graph all three functions, because then we can tell if the individual planes are approaching the airport or have passed by the airport at a particular time $t = X$ by looking at the signs of $Y1$ and $Y2$ for that value of time $t = X$. It is easy to identify the graph of $Y3$ from the other two, but it is not so easy to distinguish the graphs of $Y1$, and $Y2$. For this reason, we display the graph of $Y1$ using a bold curve. (See page 25). The screen shown is produced under ZoomFit by setting $X_{min} = 0$ and $X_{max} = 0.25$.

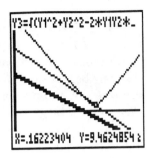

It appears that the minimum distance occurs when t is roughly $t = X = .16$ (hours). Looking at the graph, we see that the passenger jet reaches the airport first. At the moment of closest approach, the passenger jet is in a departure pattern and the fighter jet is in an approach pattern. The minimum distance appears to somewhere in the neighborhood of a sparce 10 miles. Certainly more care is required! What should we do?

This is where the more rigorous methods of calculus can play an important role. The low point will occur at a point where $c'(t) = 0$. We could press [MATH] [0] to bring up the SOLVE menu and enter $eqn : 0 = nDeriv(Y3, X, X)$ on the line $eqn : 0 =$, but our equation would be more reliable if we computed a formula for $c'(t)$ instead. **This can be done with a minimal of keyboard activity.** If $p(x)$ is any function of the form $p(x) = \sqrt{q(x)}$, then

$$p'(x) = \frac{1}{2\sqrt{q(x)}} q'(x) = \frac{1}{2p(x)} q'(x)$$

It is (relatively) easy to enter $c'(t)$ in this form. We enter

$Y4 = 1/(2 * Y3) * (2 * Y1 * -547 + 2 * Y2 * -714 - 2 * S * (Y1 * -714 + Y2 * -547)).$

Not having parentheses around the negative numbers (-547 and -714) looks confusing in print, but negative numbers produced by using the [(−)] key do not require parentheses. Look at this formula for $Y4$ and notice how the *Product Rule* and *Chain Rule* were used to produce the output.

To get a close look at $c'(t) = Y4$, deactivate $Y1, Y2, Y3$, and graph $Y4$. The result should support our initial finding that $c'(t) = 0$ near $t = X = 0.16$ hours. Press [MATH] [0] to bring up the SOLVE menu, enter $Y4 = 0$, and $X = .16$ to get $X = .16014437575894$. Certainly, **not all of these digits are significant**, but we leave X fixed at this value. Press [VARS] [▶] [1] [1] [ENTER] to get a value of -12.19897354 for $Y1$. In a similar way, get values of 11.65691571 for $Y2$ and 9.142979148 for $Y3$.

Expressing our answers with the accuracy of (pick a number) four decimal places to the right of the decimal point, it follows that the planes will arrive at their closest distance of 9.1430 miles when $t = 0.1601$ hours. At that moment, the passenger jet will be 12.1990 miles north of the airport, and the fighter jet will be 11.6570 miles northeast on the airport.

The derivative $c'(t) = \frac{dc}{dt}$ also measures the rate of change in the distance between the planes at time t. As the planes approach, this distance is **rapidly decreasing**, so we should **expect large negative values for $c'(t)$**. The rate of change will certainly be 0 at the moment of closest approach, and the distance will **rapidly increase** thereafter, so we should then **expect large positive values for $c'(t)$**.

Leave X at the moment of closest approach, where it has been through the last evaluation process. Leave $Y1$, $Y2$, $Y3$ deactivated so they don't appear in our table. Press [2ND] [TBLSET], move the cursor to TbsStart and press [X,T,θ,n]. Move the cursor to ΔTbl and enter 10/3600 (this is 10 seconds expressed in hours). Press [2ND] [TABLE] and then ▲ ▲ ▲ centering the time $t = X$ to see the screen shown. ■

3.6 Exercise Set

By using more of the available functions $Y1, Y2, \ldots, Y9, Y0$, excessive keyboard activity can often be avoided, and you may be less likely to make input mistakes. For example, if you wish to enter a function of the form $f(x) = x\sin(u(x))$ and its derivative $f'(x)$, keyboard activity can be quite excessive, especially if $u(x)$ is somewhat complicated. Instead of entering $f(x)$ and $f'(x)$ as $Y1$ and $Y2$, first set up some preliminary expressions. With $Y1 = u(X)$ and $Y2 = u'(X)$, we can enter $f(x)$ and $f'(x)$ in the form $Y3 = X\sin(Y1)$, and $Y4 = -sin(Y1) + x\cos(Y1)Y2$.

In problems 1 through 8, graph the given function. Determine the intervals where the function is increasing, decreasing, concave up, and concave down. Determine vertical, horizontal, and slant asymptotes, if they exist. Determine the order of growth, if it is appropriate. Identify points where f is not once or twice differentiable. Find the coordinates of the local maximums, local minimums, and points of inflection. Remember that graphs can have hidden features in any window, and interesting features outside of any window. Use calculus to make your arguments rigorous.

1. $f(x) = x^3 + 66x^2 - 429x - 308$

2. $f(x) = x^4 - 50x^3 - 28x^2 - 280x - 642$

3. $f(x) = x^5 + x^4 - 70x^3 + 40x^2 - 895x + 12$

4. $f(x) = x^5 - 100x^4 + 2860x^3 - 12675x^2 - 16428x + 413$

5. $f(x) = \frac{3x^2 + 6x + 19}{2x^2 + 10}$

6. $f(x) = \frac{x^3 + 3x^2 - 8x + 9}{x^2 - 5x - 4}$

7. $f(x) = |x + 4| + |x - 7 - |x - 14|| + 4$

8. $x^2(x + 9)^{1/3} + x^3 + 7$

3.6. EXERCISE SET

9. Demonstrate graphically the meaning of the *Mean Value Theorem* for
$$f(x) = \frac{x^3 - 3x^2 + 2}{x^2 + 14} \text{ on the interval } [-5, 5].$$
Use the SOLVE menu to find all of the points $x = c$ which satisfy the *Mean Value Theorem*.

10. Find a function $f(x)$ which is continuous on an interval $[a, b]$, differentiable everywhere in (a, b) except at one point, and which does not satisfy the conclusion of the *Mean Value Theorem*. Demonstrate your result with a graph.

11. Use *Newton's Method* to find the solutions to the equation
$$x \sin\left(\frac{x+1}{x^2+1}\right) = 1.$$
Specify the formulas used to determine the sequence of points which converges to a solution. Given a start up value x_0, describe graphically how x_1 and x_2 are determined. Compare your solutions to the equation with the solutions obtained using the SOLVE menu.

Newton's Method can fail if the wrong start up value is used. By looking at the graph of the function used to solve the equation, classify all start up values that would lead to failure. For each type of failure, describe graphically how the failure would occur.

12. A computer company can expect to sell
$$q = 1080 - \frac{1}{9375}p^2 + \frac{8}{75}p \quad (1000 \le p \le 3500)$$
computers per day if the selling price of a computer is p in dollars What should the selling price be in order to maximize total revenue from sales? How many computers can it expect to sell at that price? What will the revenue be from sales? Suppose that the total cost (overhead, labor and material) of manufacturing q computers per day is
$$C = 1470 + 1300q + (q - 1000)^2.$$
What should the selling price be in order to maximize profit? How many computers can it expect to sell at that price? What will the profit be?

13. A building has two very long hallways which are perpendicular to each other, and which meet at a corner of the building. One hallway is 11 foot wide, and the other is 17 foot wide. A very thin straight pole is carried down one hallway. It must make the turn so that it can be carried down the other hallway. If the pole must be kept in a horizontal position, what is the length of the longest pole that can make the turn? If the ceilings in the building are 10 feet tall, and the pole does not have to be kept horizontal, then what is the length of the longest pole that can make the turn?

14. A cylindrical storage tank is to be built to hold one million cubic feet of fuel oil. Naturally, because of structural complications, the cost <u>per square foot</u> of building

the top increases as the radius r increases, and the cost per square foot of building the sides increase as the height h increases. Suppose that these costs in dollars per square foot are:

$$C_{top} = 30 + 0.18r^{\frac{3}{2}}, \quad C_{sides} = 50 + 1.4h^2.$$

Determine the dimensions $r = r_0$ and $h = h_0$ that will minimize the cost of building the tank. Determine The total cost for these values of r and h. Compare the minimal cost to what it would be if the radius r was selected instead to be $r = r_0 \pm 10$.

15. A swimming, running, biking contest is to take place in a region described by the map shown in the figure. Map coordinates are in kilometers. The shoreline is defined by the curve $y = \frac{x^2}{10} + 2x^{\frac{1}{3}}$. A road defined by the equation $x = 3$ runs south from the shore line. The contest begins on a small island out in the lake located at the point with map coordinates (-3,3). Contestants must swim to a point of their own choosing on the shoreline. From that point, they must run due east to the road, where their assistant will be ready with a bicycle. The race continues by bicycle down the road to the point with map coordinates (3,-10) at the bottom of the map. A contestant could, for example, avoid running altogether by swimming for the point where the road meets the shoreline.

i) Ralph can swim at $3.9\,km/h$, run at $12.2\,km/h$, and bike at $34.4\,km/h$. At what position on the shoreline should he swim for in order to turn in his best personal performance?

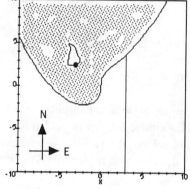

ii) Observe how Ralph's race time depends on the point (x, y) he chooses on the shoreline, by creating a table of race times centered around his best choice for x.

iii) This problem can be done in a more general way. Let S, R, and B represent the swimming, running, and biking velocities, respectively, of any contestant in the race. The appropriate expressions can be entered on your calculator in terms of the letters, S, R, and B, and then the choice of the best shoreline point to swim for can be computed for a variety of values of these constants.

Chapter 4

Integration

4.1 Area and the Definite Integral

The definition of a definite integral as the limit of its Riemann sums is so important to mathematics and its applications that we begin this chapter by using our calculator to gain some intuitive insight into this concept. Even if we downplay the many mathematical reasons for understanding this definition thoroughly, we are still left with some very important applied reasons.

As long as it is known that the integral of f on $[a, b]$ exists, its value can be determined by looking at a restricted, more manageable class of Riemann sums. In this case, if
$$a = x_0 < x_1 < \cdots < x_n = b$$
is the partition obtained by dividing $[a, b]$ into n subintervals of equal length
$$\Delta x = \frac{b-a}{n},$$
and if
$$x_0 \leq c_1 \leq x_1, x_1 \leq c_2 \leq x_2, \ldots, x_{n-1} \leq c_n \leq x_n,$$
then
$$\int_a^b f(x)dx = \lim_{n \to \infty} \sum_{k=1}^n f(c_k)\Delta x. \tag{4.1}$$

Typically, the points $c_j (j = 1, 2, \ldots, n)$ are chosen to be left end points, right end points, or mid points of their respective subintervals, and we shall call a Riemann sum obtained in this way (with this partition) **left, right, or middle Riemann sum of order n corresponding to the integral involved.** Any scheme for determining the points $c_j (j = 1, 2, \ldots, n)$, however, is acceptable. Furthermore, if f is continuous on $[a, b]$, then it is integrable on $[a, b]$, and so this simplified setting is usually sufficient. Applied problems, in particular, are usually continuous in nature.

Looking at this formula, we can see that the symbol $\int_a^b f(x)\, dx$ is really a collection of symbols, each with its own inferred meaning. Think of x ($a \leq x \leq b$) as the location of an infinitesimally thin rectangle. The symbol dx represents the width of the rectangle, and $f(x)$ represents its height or the negative of its height. The term $f(x)\,dx$ represents the ±area of the thin rectangle (positive if the rectangle is above the x-axis, and negative if the rectangle is below the x-axis).

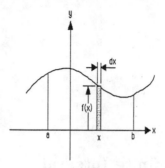

Finally, the symbol $\int_a^b$ is meant to suggest a summing process, so that $\int_a^b f(x)\,dx$ represents the sum of all the ±areas of all the thin rectangles, as the rectangles move from location $x = a$ to location $x = b$.

Frequently when problems from the real world are formulated into mathematics, they take the form of the right hand side of (4.1). A complicated problem might not be solvable "as a whole," but if we partition the problem up into a large number of parts, we might conclude that the "whole" problem is just the sum of its parts. We might be able to easily estimate each of the small parts, to get approximations $P_1, P_2, \ldots P_n$. Then a good approximation for the whole problem would be

$$\sum_{k=1}^n P_k.$$

If the approximations get better as n (the number of subdivisions) gets larger, then the answer to the original problem would be

$$Answer = \lim_{n \to \infty} \sum_{k=1}^n P_k. \qquad (4.2)$$

Without a knowledge of the definition of a definite integral, we would be left with a formidable problem. Frequently, however, (4.2) can be made to look like (4.1), so by using the definition of an integral, we can identify our problem as an integral. At this point, we have a well-understood problem, which can easily be evaluated.

Not surprisingly, there is a command on our calculator that we can use to approximate the value of an integral. It is easy to see how such a command could be created. Using formula (4.1), the command would simply evaluate a certain Riemann sum where n would be determined (by the command itself) to be big enough to produce a sufficient degree of accuracy. We could easily create such a command ourselves, and it would work reasonably well. Our calculator, it turns out, uses a similar but much more efficient process.

4.1. AREA AND THE DEFINITE INTEGRAL

We will be making frequent use of this command in the future, but not at the present time. Using this command would give us the (approximate) value of an integral, but it would not help us understand what the value means. To understand this definition, we need to compute Riemann sums directly, and our calculator is well suited to to perform this task. With a little effort on this matter we should be better equipped to translate applied problems into integrals.

There are several technological issues to consider first. This would be a good time to read Chapter 11: Lists in the TI-83 Guidebook (★86★ Chapter 11 in the TI-86 Guidebook ★86★).

A list is created on our calculator by using the left ({) and right (}) curly bracket (set bracket) symbols. To enter the list $\{3, 8, 2\}$, go to the home screen and press [2ND] [{] [3] [,] [8] [,] [2] [2ND] [}] [ENTER] (★86★ the left ({) and right (}) curly bracket symbols appear in the LIST menu—press [2ND] [LIST] ★86★). A list can be stored as a letter in the same way that a number is stored as a letter. **A number can only be stored to a single letter, but a list can be stored to a more expansive word.** (★86★ Words can be used for any storage. ★86★) To enter the above list as the letter A (having just entered the list itself), press [2ND] [ANS] (to bring the list back onto the active input line), and then [STO▶] [ALPHA] [A] [ENTER].

To use the set A in a calculation, we must do more than just press [ALPHA] [A]. (★86★ Disregard this paragraph. Using a list is easier on this machine. Just press [ALPHA] [A] [ENTER] as it normally would be entered. ★86★) **A small capital letter L must be attached as a prefix to the letter A to mark it as a list.** This prefix L appears at the very bottom (item #B) of the OPS submenu of the LIST menu. To enter the list A into another command, or somewhere else on the home screen, press [2ND] [LIST] ▶ (to get the OPS submenu), press [ALPHA] [B] to enter item #B (or press ▼ several times to scroll down to item #B), and then press [ALPHA] [A] [ENTER].

There is, however, an easier way to enter the previously defined list A into a calculation. The NAMES submenu is listed first in the LIST menu, and our set A appears in this submenu. To enter the list A into a calculation or onto the home screen, press, from the home screen, [2ND] [LIST] ▼(scroll down to A), [ENTER]. Notice how the symbol LA appears on the home screen. Continue with the calculation or press [ENTER] again to see the value of A.

Another way to avoid awkward keyboard activity is to use special names for lists that are reserved on our calculator strictly for lists. (★86★ this device has no special names for lists since any name can be used in a easy way from the keyboard ★86★). The names L1,..., L6 appear on the keyboard, and are reserved strictly for lists. To see what the current value of L1 is, simply press [2ND] [L1] [ENTER] . (An error message is returned if L1 has no value.) We do not have to use the LIST menu to insert the prefix L to these names before we use them, so these names are more convenient. With the calculator's default settings, the names L1, ..., L6 can be viewed and edited in a convenient way. Press [STAT] ,stay in the EDIT submenu, and press [1] to go to the Edit... screen. From this screen the lists L1,..., L6 can be viewed and edited. **Actually, any list can be placed on this screen,**

and the lists L1, ..., L6 can be removed. Press $\boxed{\text{STAT}}$ $\boxed{5}$ (to choose SetUpEditor) ($\star$86$\star$ Press $\boxed{\text{2ND}}$ $[\![\text{STAT}]\!]$ $\boxed{\text{F2}}$ (to select the EDIT submenu) $\boxed{\text{F5}}$ (to select the OPS submenu) $\boxed{\text{MORE}}$ $\boxed{\text{MORE}}$ $\boxed{\text{MORE}}$ $\boxed{\text{F3}}$ (to select SetLE), enter the names of the lists you wish to appear on this screeen, separated by commas, and press $\boxed{\text{ENTER}}$. The calculator responds with "done", removes all previous lists from the STAT EDIT screen and places the desired lists on the screen in the same order as they were listed in the SetUpEditor ($\star$86$\star$ SetLE $\star$86$\star$) statement.

Most commands on our calculator that ordinarily act on real or complex numbers have the very useful property of being listable. To explain this important term, let $f(x)$ be a function of a real variable x, and let A be a list of real numbers. We say $f(x)$ is listable, if $f(A)$ is the list of all elements $f(x)$ as x ranges over A. In the same way, we say that the operator $(+)$, for example, is listable, if $A + 10$ is the list of all elements of the form $x + 10$ as x ranges over A.

Suppose that we use $A = \{3, 8, 2\}$ (entered into our calculator earlier), and enter $Y1 = X \wedge 2$ in the Y= menu. Here are some calculations, we can perform using the listable property of various commands.

$$Y1(\text{L}A) = \{9, 64, 4\}, \qquad .2 * \text{L}A + 3 = \{3.6, 4.6, 3.4\}$$
$$\cos(\text{L}A * \pi) = \{-1, 1, 1\} \qquad 72/\text{L}A = \{24, 9, 36\}.$$

When a list is small enough, its values can be entered directly between the curly bracket symbols as we did when we defined the list A above. This approach, however, is not very convenient when the list is large. As long as the elements in a list can be determined by some formula, **the list can be created using the sequence command**. This command is the 5th item in the OPS submenu of the LIST menu. Its first argument is an expression $F(B)$ in some variable B, The second entry specifies the variable. The third and forth entries are the starting and ending values of the variable B.

$$seq(F(B), B, j, n) = \{F(j), F(j+1), F(j+2), \ldots, F(n)\}.$$

The sequence command can be given an optional fifth argument ΔB which specifies an increment on the variable B. If k is the largest integer with $j + k\Delta B \leq n$, then

$$seq(F(B), B, j, n, \Delta B) = \{F(j), F(j+\Delta B), F(j+2\Delta B), \ldots, F(j+k\Delta B)\}.$$

A list on our calculator can contain no more than 999 elements ($\star$86$\star$ 3000 elements or more depending on the amount of available memory $\star$86$\star$). Since lists can take up a great deal of storage space when they are stored in memory (given names), they should be removed when they are no longer needed. Press $\boxed{\text{2ND}}$ $[\![\text{MEM}]\!]$ $\boxed{2}$ $\boxed{4}$, select the set which is to be deleted and press $\boxed{\text{ENTER}}$.

These tools can be used very effectively to create Riemann sums. Unfortunately, our calculator cannot draw (without excessive labor) the complex of rectangles associated with a Riemann sum. We can, however, use our calculator to graph the function, and with a graph copied on paper, the rectangles involved can easily be penciled in. This picture should accompany our work.

4.1. AREA AND THE DEFINITE INTEGRAL

Example 4.1 *Estimate the value of $\int_2^8 (x^2 + 3)\, dx$ by computing its left, right, and middle Riemann sums of order n for $n = 50$ and $n = 200$. Evaluate the integral using the calculator's numerical integration command.*

Enter $f(x) = x^2 + 3$ as Y1 in the Y= menu. We will let $L(n), R(n), M(n)$ denote the left, right, and middle Riemann sum respectively of order n for this integral.

It would help to motivate our work, and to understand what these sums mean if we drew the rectangular complexes, but this would be tedious to do for $n = 50$. It might not even achieve the desired result. The rectangles would be so thin, that it would be hard to distinguish between them. We show instead, from left to right in the figure below, the rectangular complexes corresponding to the sums $L(10), R(10)$, and $M(10)$. They were produced for this manual with the computer program Maple, but there is no need for such a careful drawing. We could always graph the function with our calculator, and from the graph, the rectangles can just as easily be drawn by hand on paper. From the figures, it is clear that for this function and any n, $L(n)$ will be an under approximation, $R(n)$ will be an over approximation, and $M(n)$ will be the best approximation of the three.

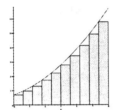  

If the interval [2,8] is partitioned into n subintervals of equal length, then each subinterval has length $\Delta x = (8-2)/n = 6/n$, and the partition points are

$$2 = 2 + 0\frac{6}{n} < 2 + \frac{6}{n} < 2 + 2\frac{6}{n} < 2 + 3\frac{6}{n} < \ldots < 2 + n\frac{6}{n} = 8.$$

The left end points would be this display with the last point removed, and the right end points would be the same display with the first point removed. The midpoints can be obtained by adding $\frac{\Delta x}{2} = \frac{3}{n}$ to the left partition points. This produces

$$2 + \frac{3}{n} = 2 + \frac{3}{n} + 0\frac{6}{n} < 2 + \frac{3}{n} + 1\frac{6}{n} < 2 + \frac{3}{n} + 2\frac{6}{n} < \ldots < 2 + \frac{3}{n} + (n-1)\frac{6}{n} = 8.$$

It would be convenient to use the letter X as the variable for our sequence, since it can be entered by just pressing $\boxed{\text{X,T,}\theta\text{,}n}$. We shall, however, use the letter J instead, just to avoid confusing our sequence variable with the "idea" of x as the variable of integration.

From the home screen, press $\boxed{\text{2ND}}$ $\boxed{\text{LIST}}$ ▶ $\boxed{5}$ to enter the sequence command onto the active line of the home screen. Then, with appropriate keyboard activity, enter $seq(2 + J * 6/50, J, 0, 49)$ and press [STO▶] [ALPHA] ([P]) . The list P is the sequence of left end points.

We will now use the listable properties of our calculator. Enter $Y1(\text{L}P) * 6/50$ with appropriate keyboard activity and press [STO▶] [ALPHA] ([A]), to get the list A of all the rectangular areas for our left sum. Notice the prefix L that appears in from of the list P. The list P is entered through the NAMES submenu of the LIST menu. (★86★ There is no prefix L, and P can be entered directly from the keyboard, or from the NAMES submenu of the LIST menu. ★86★) All that is left to do to produce a value for $L(50)$ is add up the elements in A. To do this **we use the sum command which is the fifth element in the MATH submenu of the LIST menu.** (★86★ The sum command is in the OPS submenu of the LIST menu—press [2ND] [LIST] [F5] [MORE] [F1] ★86★) Press [2ND] [LIST] ▶ ▶ [5] to enter the sum command onto the active line of the home screen, and then press [2ND] [LIST] ▼(scroll down to A) [ENTER] [ )] [ENTER] to get $L(50) = 182.4144$.

It is easy to compute the list of midpoints for $n = 50$ from the list P of left hand points. Taking advantage of listable properties, the midpoint list would be $3/50 + \text{L}P$. The middle sum $M(50)$ can be computed in one step by entering $sum(Y1(3/50 + \text{L}P) * 6/50)$ to get $M(50) = 185.9928$.

In order to get $R(50)$ from our previous work, we must edit the formula for P. Pressing the keys [2ND] [[ENTRY]] **returns the previous input statement to the active input line. By pressing this combination repeatedly, we can return input statements made several steps back to the active line on the home screen where they can be easily edited**. As soon as the statement $seq(2 + J * 6/50, J, 0, 49) \to P$ appears we can change the beginning and end values of J from 0,49 to 1,50. We choose instead to compute $R(50)$ directly, in one step, by entering $sum(seq(Y1(2 + J * 6/50) * 6/50, J, 1, 50))$ to get $R(50) = 189.6144$.

To evaluate the Riemann sums of order $n = 200$ we will take advantage of the complete input statement we just entered for $R(50)$. Pressing [2ND] [[ENTRY]] returns the last input statement to the active input line. Change the number 50 to 200 in three locations (Press [2ND] [[INS]] at the location of 50 to make room for the extra digits) to get $sum(seq(Y1(2 + J * 6/200) * 6/200, J, 1, 200))$. Press [ENTER] to get $R(200) = 186.9009$. Press [2ND] [[ENTRY]] again to return the last input statement to the active input line. Change the beginning and ending numbers on J from 1, 200 to 0, 199 to get $sum(seq(Y1(2 + J * 6/200) * 6/200, J, 0, 199))$. Press [ENTER] to get $L(200) = 185.1009$. Press [2ND] [[ENTRY]] one more time and edit the formula to get the middle sum. This time we add $\frac{\Delta x}{2} = 3/200$ to the left end points to get $sum(seq(Y1(2 + 3/200 + J * 6/200) * 6/200, J, 0, 199))$. Press [ENTER] to get $M(200) = 185.99955$.

From these calculations, and from the above figures, we have (Inequalities like the following only hold for functions that are increasing over the interval of integration.)

$$L(50) < L(200) = 185.1009 < \int_2^8 (x^2 + 3)\, dx < 186.9009 = R(200) < R(5).$$

$$\int_2^8 (x^2 + 3)\, dx \approx M(200) = 185.99955$$

The drawing for $M(10)$ suggests that the value of $M(200)$ is the best of our approximations. Finally, we can see that the values involved support the notion that

$$\lim_{n\to\infty} L(n) = \lim_{n\to\infty} R(n) = \lim_{n\to\infty} M(n) = \int_2^8 (x^2+3)\,dx.$$

To compute the numerical value of an integral directly on our calculator we use the *fnInt* command, item #9 on the MATH submenu of the MATH menu. (★86★ The command *fnint* is in the CALC menu–press [2ND] [CALC] [F5] .∴ ★86★) Press [MATH] [9], enter the input statement $fnInt(Y1, X, 2, 8)$, and press [ENTER], to get $\int_2^8 (x^2+3)\,dx \approx 186$. ■

As we will discover in the next section, the value of 186 that we obtained with the *fnInt* command also happens to be an <u>exact</u> value for this integral. Our calculator's numerical integration command is not always so accurate, but its accuracy is naturally a matter that deserves our consideration. The command *fnInt* has an optional fifth argument which controls the accuracy of the computation. Given a small number ϵ, the input statement $fnInt(f(X), X, a, b, \epsilon)$ computes the integral with an accuracy

$$\left| \int_a^b f(x)\,dx - fnInt(f(X), X, a, b, \epsilon) \right| \leq \epsilon.$$

When the fifth argument ϵ is not specified, a value of $\epsilon = 10^{-5}$ is assumed.

There is another integration command which is more graphical. The command $\int f(x)dx$ is item #7 in the CALC menu (★86★ $\int f(x)$ is in the MATH submenu of the GRAPH menu—press [GRAPH] [MORE] [F1] [F3] ★86★). Using this command will shade in the "area-like" region involved in the integral, and produce a value for the integral. Its accuracy is limited somewhat by how accurately one can select pixels which correspond to the end points of the interval of integration.

4.2 The Fundamental Theorem of Calculus

According to the *Fundamental theorem of calculus*, if f is continuous on the interval $[a, b]$, then the function F defined by

$$F(x) = \int_a^x f(t)dt, \tag{4.3}$$

is differentiable on $[a, b]$, and $F'(x) = f(x)$ for all x in $[a, b]$. We used t instead of x as the variable of integration in the formula so that the letter would not conflict with the x used as an upper limit of integration. The choice of letter used as a variable of integration has no bearing on the value of a definite integral. Any two antiderivatives differ by a constant. Because of this, an antiderivative of a given function $f(x)$ is uniquely determined as soon as we specify a value at one point. Notice that the antiderivative $F(x)$ defined by Formula (4.3) has the value $F(a) = 0$.

As a consequence of (4.3), if G is any antiderivative of $f(x)$, then

$$\int_a^b f(x)dx = G(b) - G(a) = G(x)\,|_a^b\,. \tag{4.4}$$

Formula (4.3) is important for reasons other than the fact that it gives us formula 4.4, which is the most basic formula for evaluating definite integrals. Formula 4.3 also implies that every continuous function has an antiderivative, and it gives us a formula (sort of) for it. This is of considerable consequence in its own right, and so we begin by using our calculator to study F.

Example 4.2 *Let $f(x) = x^2 + 3$ be the function from our previous example. Use the calculator's integration command to define the function*

$$F(x) = \int_2^x (t^2 + 3)\,dt.$$

Notice that its value at $x = 2$ is $F(2) = 0$. Find, analytically, a formula for the function $G(x)$ such that $G'(x) = f(x)$ and $G(2) = 0$. Compare the values of the functions $F(x)$ and $G(x)$ by creating a table of values over the interval $0 \leq x \leq 10$. (Recall, from the last example, that $F(8) = 186$) Explain why $F(x)$ is negative for $x < 2$.

Enter the function $f(x)$ as $Y1$ in the Y= menu, and enter $fnInt(Y1, X, 2, X)$ as Y2. This is a valid way to enter the expression for $F(x)$. Our calculator knows that the $X's$ used in the second and forth arguments play different roles.

As a point of interest however, we could have used a different variable of integration. In order to avoid confusion over the meaning of $F(x)$, let us enter these formulas with T as the variable of integration. Enter $Y1 = T \wedge 2 + 3$ and $Y2 = fnInt(Y1, T, 2, X)$. Which ever method is used, deactivate $Y1$ so that its values do not appear in our table.

The basic antiderivative of $f(x)$ is $G(x) = \frac{x^3}{3} + 3x + C$, for some constant C. We could use our calculator to evaluate C but it is simple enough to do by hand. We want $G(2) = 0$, and the equation $G(2) = \frac{8}{3} + 6 + C = 0$ gives $C = -26/3$. With this value for C, enter the expression $G(x)$ as $Y3$ in the Y= menu.

We are ready to compare the values of $F(x)$ and $G(x)$. Press [2ND] [TBLSET], set TblStart= 0, ΔTbl= 1, and press [2ND] [TABLE] to see the table shown. Notice how the values for $F(x)$ and $G(x)$ are the same for every entry in the table. To see the values of $F(x)$ and $G(x)$ for values of x that are not displayed on the screen, use the ▲ and ▼ keys to scroll up or down on the table. In particular, notice that $F(8) = G(8) = 186$, the value for the integral that we computed in the last example.

Since $f(x) = x^2 + 3$ is always positive, the value of $\int_a^b f(x)\,dx$ will be positive whenever $a < b$, because the integral represents the area under the curve. In our table, the value, for example, of

$$F(0) = \int_2^0 f(x)\,dx = -\int_0^2 f(x)\,dx$$

must, therefore, be negative. ∎

This example certainly supports the idea that Formula (4.3) defines an antiderivative of $f(x)$, and also the particular one whose value at $x = a$ is $F(a) = 0$.

In the next example, we show another way to drive home the point that Formula (4.3) defines an antiderivative of $f(x)$.

Example 4.3 *Define the functions $f(x)$ and $F(x)$ by*

$$f(x) = x\cos(x) \quad and \quad F(x) = \int_0^x f(x)\,dx.$$

Show that $F(x)$ is an antiderivative of $f(x)$ by creating a table which shows that $F'(x) = f(x)$ for a range of values of x.

To do this on our calculator, **we must enter $Y1 = X*cos(X)$ as a function of X and not as a function of T**. Enter $F(x)$ in the form $Y2 = fnInt(Y1, X, 0, X)$, using an X in both the second and forth arguments, and then deactivate $Y2$ so that its values do not appear in our table. Finally, enter $nDeriv(Y2, X, X)$ as $Y3$. Set TblStart$= 0$, ΔTbl$= 1$, and press [2ND] [TABLE] to see the values of $f(x)$ and $F'(x)$ for a range of values of x. Their agreement everywhere is more evidence in support of the *Fundamental Theorem of Calculus*. ∎

X	Y1	Y3
0	0	0
1	.5403	.5403
2	-.8323	-.8323
3	-2.97	-2.97
4	-2.615	-2.615
5	1.4183	1.4183
6	5.761	5.761

X=0

4.3 Some Theorems of Integration

One of the most important tools we have in mathematics for manipulating and evaluating integrals analytically is *Integration by Substitution*, which states that

$$\int_a^b f(g(x))g'(x)\,dx = \int_{g(a)}^{g(b)} f(u)\,du.$$

This result, is not needed to evaluate an integral on our calculator, but it is still an important part of the exact evaluation of an integral.

Our calculator can, however, be used to gain some insight into this important result.

Example 4.4 *Let $f(x) = \sqrt{x}$, and $g(x) = 4 + 2\sin(x)$. Show that*

$$\int_{-\pi/2}^{\pi/2} f(g(x))g'(x)\,dx = \int_{g(-\pi/2)}^{g(\pi/2)} f(u)\,du,$$

by evaluating both sides in the following way. Graph each integrand on an appropriate window. Use windows with the same scale so that the regions involved can be compared. Do their $\pm$areas look the same? Compute $M(10)$, the so-called middle sum for each integral. Evaluate each integral using the fnInt command.

We apply the *Integration by Substitution Theorem* with $u = 4 + \sin(x)$ to get the relationship

$$\int_{-\pi/2}^{\pi/2} \sqrt{4+\sin(x)}\, 2\cos(x)\, dx = \int_2^6 \sqrt{u}\, du$$

that we wish to establish graphically and numerically.

Enter $f(g(x))g'(x) = \sqrt{4+\sin(x)}\cos(x)$ as $Y1$ and $f(u) = \sqrt{u}$ as $Y2$ (entered, of course, in the form $Y2 = \sqrt{(X)}$). To compare the "area-like" regions corresponding to the graphs of $Y1$ and $Y2$, we want the the horizontal scales to be the same length for both graphs, and we want the vertical scales to be the same for both. For that reason, we graph $Y1$ on the interval $[-2, 2]$ (rather than $[\pi/2, \pi/2]$) and $Y2$ on the interval $[2, , , 6]$, both of which are intervals of length 4. The same vertical range of $Y_{min} = 0$ and $Y_{max} = 5$ was used to produce the screens shown.

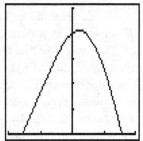

It is only a minor point, but for what it is worth, the two regions appear (visually) to have areas which might be the same. Each viewing screen has dimensions of 4 units by 5 units, and the regions appear to occupy about two fifths of the window. That would mean that the integrals should evaluate to somewhere in the neighborhood of $4 \times 5 \times 2/5 = 8 units^2$. **This visual and very imprecise analysis is not without merit.** It is very easy to press the wrong key, or to make some other error when technology is being used. We need to stay connected to the output of our work—sometimes in just simple visual ways like this—in order to make sound judgments about validity of our results. **A good scientist should be a skeptic! Our minds should always be open to the potential for error**.

Well! Now let us be more precise. For the first integral over the interval $[-\pi/1, \pi/2]$, the width of each subinterval is $\Delta x = \pi/20$, and for the integral over $[2, 6]$ it is $\Delta x = 4/20 = 1/5$. We compute the sum $M(20)$ in one step by entering

$$sum(seq(Y1(-\pi/2 + \pi/40 + X * \pi/20) * \pi/20, X, 0, 19))$$

to get an approximate value of 7.920289 for the left hand integral. **The potential for error is quite large when this statement is entered**. It helps to know that the answer should be approximately 8. For the integral on the right, we enter

$$sum(seq(Y2(2 + 1/10 + X * (1/5)) * (1/5), X, 0, 19))$$

to get an approximate value of 7.912589. (We multiplied by $(1/5)$ rather than divide by 5 only to show how $\Delta x = 1/5$ fits into the sum.)

For more accuracy we compute $fnInt(Y1, X, \pi/2, \pi/2)$ to get 7.912340888 for the left hand integral, and we enter $fnInt(Y2, X, 2, 6)$ to get 7.912340888. This is certainly strong numerical evidence that the integrals are equivalent. ∎

4.4. NUMERICAL INTEGRATION

The *Mean Value Theorem for Integrals* asserts the existence, for a continuous function $f(x)$ on $[a, b]$ of a point c in the interval $[a, b]$ such that

$$\int_a^b f(x)\, dx = f(c)(b - a).$$

The right hand side of this equality can be thought of as the $\pm$area of a rectangle of $\pm$height $f(c)$ and of horizontal width $(b - a)$.

Example 4.5 *Graph the function $f(x) = \frac{x}{x-4}$ on the interval $[0, 3]$ and find visually a point c in $[0, 3]$ for which $\int_0^3 f(x)\, dx = f(c)(3 - 0)$. Use the DRAW menu to draw the corresponding rectangle. Find a precise numerical value for c by using the SOLVER menu.*

We enter $f(x)$ as $Y1$ in the Y= menu, set $X_{min} = 0$, $X_{max} = 3$ and graph the negatively valued $f(x)$ under ZoomFit. Press [TRACE] and move the trace cursor to the point on the graph where the bottom edge of the sought after rectangle might be located. Press [2ND] [DRAW] [3] to enter a horizontal line through the point where the trace cursor is located ($\star$86$\star$ the DRAW submenu is in the GRAPH menu $\star$86$\star$). The x-coordinate $X = 1.9148936$ is the approximate value of c. The $\pm$height of the rectangle is the y-coordinate $Y = -.9193548$ of this point on the graph. If we multiply this number by 3, the width of the interval, we should get an approximate value for the integral. The easiest way to enter this product is to go to the home screen (without moving the trace cursor) and press [VARS] [▶] [1] [1] [×] [3] [ENTER], to get an approximate value of -2.755102041 for the integral. To see how close this value of c is to the correct answer, we enter $fnInt(Y1, X, 0, 3)$ to get a value of -2.545177444 for the integral. This compares favorably to the negative area of the rectangle. For such a rough approximation, our value for c is reasonably accurate.

To compute c more accurately, we use the SOLVER menu. Press [2ND] [ANS] [STO▶] [ALPHA] ([I]) [ENTER] to store the numerical value of the integral we just computed to the letter I. Press [MATH] [0] (the tenth item) [ENTER] to get the SOLVER menu ($\star$86$\star$ press [2ND] [SOLVER] $\star$86$\star$). At "$eqn : 0 =$" enter $eqn : 0 = I - Y1(C) * (3 - 0)$ ($\star$86$\star$ enter $eqn : I = y1(C) * 3$ $\star$86$\star$). Press ▼ ▼ to bypass X, enter an approximate value of 1.9 for C and with the cursor on the line containing the unknown C, press [ALPHA] ([SOLVE]) to get a value of $c = 1.8359574386672$. We could have done this with less keyboard activity by entering $eqn : 0 = I - Y1 * 3$ and solving the equation for X. We used C instead to make the solution more readable.

4.4 Numerical Integration

The definition of a definite integral suggests several numerical techniques for approximating the value of an integral. We have already discussed and used the left, right, and middle sums, $L(n), R(n), M(n)$ to approximate an integral. Recall that when $n = 200$, our best approximation $M(200)$ gave us an answer with 6 digits of accuracy, but our calculator had to work reasonably hard to produce that answer. We can use the idea of a definite integral as a limit of Riemann sums to create much more efficient numerical techniques.

Our calculator uses fairly sophisticated techniques for approximating the value of a definite integral. Rather than discuss these techniques, we look instead at *Simpson's Rule*, one of two numerical integration techniques typically introduced in a standard calculus course. An understanding of this rule will give us some insight into the general idea of numerical integration, and, will, hopefully, give us an appreciation of our calculator's *fnInt* command in the process.

Simpson's Rule for approximating the value of the integral $\int_a^b f(x)\,dx$ begins by dividing the interval $[a, b]$ into an even number $n = 2m$ of subintervals of equal length $h = \Delta x = \frac{b-a}{n}$. If we let $y_j = f(a + jh)$ $(j = 0\ldots,n)$, then the formula for approximating the integral is

$$S(n) = h/3\,[y_0 + 4y_1 + 2y_2 + 4y_3 + 2y_4 + 4y_5 + \ldots + 2y_{n-2} + 4y_{n-1} + y_n].$$

This can be turned into an indexed sum that we can enter on our calculator in two different ways. Group the terms in $S(n)$ in the form

$$S(n) = h/3\,[y_0 + 4y_1 + (2y_2 + 4y_3) + (2y_4 + 4y_5) + \ldots + (2y_{n-2} + 4y_{n-1}) + y_n],$$

and notice that the terms $4y_j$ occur with j odd, and the terms $2y_j$ occur with j even. This leads to

$$S(n) = \frac{h}{3}\left[y_0 + 4y_1 + y_n + \sum_{j=1}^{m-1}(2y_{2j} + 4y_{2j+1})\right] \quad (n = 2m).$$

The terms in $S(n)$ can also be grouped in the form

$$S(n) = h/3\,[(y_0 + 4y_1 + y_2) + (y_2 + 4y_3 + y_4) + \ldots + (y_{n-2} + 4y_{n-1} + y_n)],$$

and this leads to the form

$$S(n) = \frac{h}{3}\left[\sum_{j=0}^{m-1}(y_{2j} + 4y_{2j+1} + y_{2j+2})\right] \quad (n = 2m).$$

If the fourth derivative f^4 is continuous on $[a, b]$, then the error

$$E(n) = \left|\int_a^b f(x)\,dx - S(n)\right|$$

satisfies

$$E \leq \frac{(b-a)^5}{180n^4}\left[max|f^{(4)}(x)|\right], a \leq x \leq b.$$

Example 4.6 *Approximate the value of $\int_0^2 x^2 sin(3x)\,dx$ by computing the middle sum $M(50)$ and the Simpson's Rule sum $S(50)$. Compare the values to the value obtained by using the calculator's numerical integration command fnInt. Determine the corresponding Simpson's Rule error. Use the error term and the value of $S(50)$ to give and upper and lower bound on the actual value of the integral.*

4.4. NUMERICAL INTEGRATION

Enter $f(x)$ as $Y1$, and set $H = \frac{b-a}{n} = \frac{2}{50}$ by entering 2/50 and pressing [STO▶] [ALPHA] ([H]) [ENTER]. To evaluate $M(50)$, we enter

$$sum(seq(Y1(0 + H/2 + J*H)*H, J, 0, 49))$$

on the home screen and press [ENTER] to get $M(50) = -1.40805592$. To evaluate $S(50)$ (remember that $m = n/2 = 25$), enter

$$H/3 * sum(seq(Y1(2*J*H) + 4*Y1((2*J+1)*H) + Y1((2*J+2)*H), J, 0, 24)))$$

on the home screen and press [STO▶] [ALPHA] ([S]) [ENTER] to get the *Simpson's Rule* approximation $S = S(50) = -1.407363121$ for the value of the integral. Enter $fnInt(Y1, X, 0, 2)$ and press [ENTER] to get our calculator's numerical approximation of -1.407362064. It is evident that $S(50)$ is more accurate that $M(50)$.

In order to compute the error term, we must find the largest value of $|f^{(4)}(x)|$ on the interval $0 \leq x \leq 2$. We can only use the numerical differentiation command $nDeriv$ twice on our calculator, so we must compute at least $f''(x)$. We choose instead to compute the fourth derivative symbolically. After some effort, we enter

$$|f^{(4)}(x)| = \left|-108\sin(3x) - 216x\cos(3x) + 81x^2\sin(3x)\right|$$

as $Y2$. The process begins by pressing [MATH] ▶ (to get the NUM submenu) [1] to enter the absolute value command *abs* onto the line $Y2 =$ of the Y= menu. The rest of this expression can then be entered in a straight forward manner.

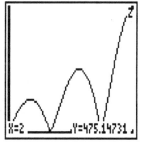

It is tempting to use the *fMax* command in the MATH menu in order to get the largest value of $Y2$, but **the *fMax* command will not give the desired answer.** This command will only give the x-coordinate of the local maximum and **not** the x-coordinate of the **absolute** maximum which is what we need. Enter $X_{min} = 0$ and $X_{max} = 2$ in the WINDOW menu, and graph $Y2$ under ZoomFit. Press [TRACE] and move the cursor to the high point on the curve to get the screen shown. Using the error term formula given above, the error $E(50)$ is

$$E(50) \leq \frac{2^5}{180 \times 50^4} \times 476 \leq 1.36 \times 10^{-5}$$

To be on the safe side, a slightly bigger value of $Y4$ was used. The answer was truncated up to a slightly larger number, because there is no real significance to an error evaluation with so many digits of accuracy. After all, what we want from an error term is how small it is. We want to know how many digits in the evaluate of $S(50)$ are correct.

To finish up, let us use the [STO▶] key to give E the value $E = 1.36 \times 10^{-5}$. The error inequality

$$\left|\int_a^b f(x)\,dx - S(50)\right| \leq E,$$

leads to

$$-1.407376721 = S(50) - E \leq \int_a^b f(x)\,dx \leq S(5) + E = -1.407349521.$$

Recall that we already stored the value of $S(50)$ to the letter S, so the left and right hand sides were easily evaluated from the home screen. Just press, for example, [ALPHA] ([S]) [−] [ALPHA] ([E]) [ENTER] for the left hand side. ∎

4.5 Exercise Set

1. Use the seq( ) and sum( ) commands to compute the following sums.
   a) $\sum_{j=1}^{200} \frac{j}{j^2+1}$ b) $\sum_{k=14}^{150} \frac{(-1)^k}{k}$ c) $\sum_{n=1}^{300}(3n^2 - 5n)$

2. Approximate the value of the integral $\int_{-3}^{2} \sqrt{x^2+4}\,dx$ by computing the following Riemann sums. In each case carefully draw the graph and the approximating rectangles on paper.   a) $L(10)$ b) $R(15)$ c) $M(5)$

3. a) Use the *fnInt* command to evaluate the integral $\int_1^4 \frac{x^2}{x+2}$. Approximate the value of the integral by computing
   b) $L(n)$ for $n = 40, 60, 80, 100$
   c) $R(n)$ for $n = 50, 100, 150, 200$
   d) $M(n)$ for $n = 16, 32, 64, 128$
   (Hint: Enter a Riemann sum formula from the keyboard only once. To bring the formula back onto the active line of the home screen so that it can be edited and evaluated again with a different value of n, press [2ND] [[ENTRY]], repeatedly if necessary until the appropriate formula appears.)

4. Let $f(x) = \frac{x}{x+\sin(x)}$
   a) Use the *Fundamental Theorem of Calculus* to define an antiderivative $F(x)$ of $f(x)$ which is defined for x in the interval $[1, \infty)$. Enter $F(x)$ as a function in the Y= menu of your calculator.
   b) Evaluate $F(18)$ and $F(3)$ as decimal numbers. Compare the value of $F(18) - F(3)$ with the value of $\int_3^{18} f(x)dx$ obtained by using the *fnInt* command. Why are the answers the same?
   c) Show that $F'(x) = f(x)$ for a range of x values by creating a table of values comparing $F'(x)$ and $f(x)$.
   d) Let $h(x) = F(g(x))$, where $g(x) = 3x^4 + 7x^2 + 9$. Compute $h'(3)$ directly from the functions $F(x)$ and $g(x)$ without entering $h(x)$ onto your calculator.
   e) Let $p(x) = G(F(x))$, where $G(x) = \frac{x^2}{\sqrt{9+x^2}}$. Compute $p'(5)$ directly from the functions $F(x)$ and $G(x)$ without entering $p(x)$ onto your calculator.

5. Evaluate the following integrals.
   a) $\int_2^{10}(2 + x + x^2)\sqrt{3x - 5}\,dx$
   b) $\int_{-\pi/6}^{\pi/12} \sin^3(2y)\cos^4(2y)\,dy$
   c) $\int_0^3 (x^3)\sqrt{25 - x^2}\,dx$

4.5. EXERCISE SET

6. a) Find a function $F(x)$ such that $F'(x) = x^2\sqrt{4+x^2}$ and $F(3) = 0$. Do not look for an algebraic formula for $F(x)$.
b) How do we know that such a function exists? What is its domain?
c) Create a table which displays the values of $F(x)$ for positive integers x. Display the table with $x = 12$ appearing as the top entry in the table.
d) Write the equation of the line tangent to the graph of $F(x)$ at the point on the graph corresponding to $x = 6$. Is the tangent line a good approximation for $F(x)$ for points x near $x = 6$? With the tangent line expressed in the form $y = L(x)$, create a table comparing the values of $F(x)$ and $L(x)$ for points x near $x = 6$.

7. Find a function $F(x)$ such that $F'(x) = \sqrt{x}\sin(x)$ and $F(7.4) = -3.8$. The function $F(x)$ is not expressible in terms of the elementary special functions of mathematics. How do we know that such a function exists? What is its domain? Create a table which displays the values of $F(x)$ for decimal numbers x expressed in tenths of units. Display the table with $x = 7.0$ appearing as the top entry in the table.

8. Show that $\int_a^b f(x)g(x)\,dx = \int_a^b f(x)\,dx \int_a^b g(x)\,dx$ is *false*, by using an appropriate counterexample.

9. a) Change the variable of integration for the integral $\int_{-2}^{1} \frac{x^2}{\sqrt{x^3+9}}\,dx$ by letting $u = x^3 + 9$.
b) Graph the region involved with each integral, using the same scale with both graphs, so that the integrals can be at least roughly compared visually. Estimate each integral roughly by looking at the graph. Do the values look the same?
c) Estimate the values more accurately by computing $M(10)$. Draw the graphs and the approximating rectangles on paper.
d) Verify, using the *fnInt* command, that the two integrals have the same value.
e) Evaluate one of the integrals exactly using analytic techniques.

10. Using very rough graphical techniques, and then again more accurately using the SOLVER menu, find all values c such that $f(c)(b-a) = \int_a^b f(x)\,dx$ for
a) $f(x) = 4 + x\cos(x)$, $a = 1$, $b = 3$,
b) $f(x) = 2\sin^3(x/5)$, $a = -6\pi$, $b = 4\pi$.

11. Evaluate $\int_1^3 \sqrt{x^2+1}\,dx$ as a decimal. Use the *Trapezoidal Rule* and *Simpson's Rule* with $n = 20$ to estimate the integral. Approximate the size of the error terms by using formulas for the error terms given in your calculus text (or this manual). Use the error terms to give upper and lower bounds on the exact value of the integral.

12. Use *Simpson's Rule* to approximate the integral $\int_{-5\pi/4}^{2\pi/3} \sqrt{x}\sin(3x)\,dx$. Find and use a value of n large enough to give 8 digits of accuracy.

Chapter 5

Applications of Integration

We began the last chapter with a remark concerning the importance of the definition of a definite integral as the limit of its Riemann sums. As you can now see in your calculus text, science is rich with examples of situations which have an interpretation as a limit of some Riemann sum, and hence as a definite integral. In this chapter, we follow up on the ideas presented in your calculus text, and show how as graphing calculator can be used to solve some interesting applied problems.

5.1 Area and Volume

Example 5.1 *Find the area of the region between the curves defined by the equations $x^2 + y^2 = 25$, and $x = 7y - 3y^2 - 1$.*

To begin this problem, a plot is essential. We could solve both equations for y in terms of x, but given the nature of the second equation, it is much more convenient to solve both equations for x in terms of y. Solving the first equation, a circle, for x produces $x = \pm\sqrt{25 - y^2}$. As we are all well aware, functions are entered into the Y= menu in the form $Y = f(X)$. To graph these functions we must first interchange x and y, and allow our imagination to interpret the vertical axis (pointing up) as the positive x-axis, and the horizontal axis (pointing to the right) as the positive y-axis. Should we instead simply rotate our calculator 90° counterclockwise? That would indeed turn the positive y-axis up, as we are accustomed to seeing it, but it would make the positive x-axis point to the left rather than the right.

Nothing is lost if we simply accept the horizontal and vertical interchange, and so we enter $Y1 = \sqrt{(25 - X \wedge 2)}, Y2 = -Y1, Y3 = 7X - 3X \wedge 2 - 1$. Focusing our attention on the circle of radius 5 centered at the origin, set $X_{min} = -6, X_{max} = 6, Y_{min} = -6, Y_{max} = 6$, and press GRAPH. The circle looks like an ellipse, because the viewing screen is not square. There is nothing wrong with accepting this and moving on to more important matters. The graph, however, can be adjusted by pressing [ZOOM] [5] (for ZSquare) to produce the screen shown. We can see that the region is bounded below by the lower half of the circle and above by the parabola.

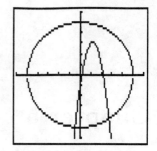

One very interesting way to avoid this horizontal-vertical problem is to define the curves parametrically as the set of all point (x, y) where x and y are expressed in the form $x = f(t), y = g(t)$, where t is some third variable, called the parameter. The parameter t is frequently thought of as $t = time$. This topic could be discussed at the present time, but it is typically postponed to a later date in the calculus sequence. Since the slight misrepresentation of the above graph is not a critical issue that needs to be fixed, we shall not bring up the matter of parametric curves until later.

Imagine a thin vertical rectangle (vertical on our calculator) drawn between the curves like the one in the left hand figure below. Its height is $Y3 - Y2$, its width is dX, and so its area is $(Y3 - Y2)dX$. To get the total area of the region, we add up (in the sense of integration) the areas of all of the "thin" rectangles as they move from the left hand corner of the region to the right hand corner. This leads to an integral of the form $\int_a^b (Y3(X) - Y2(X))\, dX$, but before we can enter it into our calculator, we must determine the X-coordinates of the points of intersection.

We can get approximate values for these points by pressing [TRACE] and moving the trace cursor to the points of intersection. This gives us approximate values of $X \approx -0.38$ and $X \approx 2.7$. To get more accurate values, we go to the SOLVER menu by pressing [MATH] [0] (the tenth item). Enter $eqn : 0 = Y2 - Y3$ at $eqn : 0 =$, press ▼, enter -.38 for X, and press [ALPHA] ([SOLVE]), to get $X = -.47253580629974$. Go to the home screen and press [X,T,θ,n] [STO▶] [ALPHA] ([A]) [ENTER] to store the value as A. We return to the SOLVER menu and repeat this procedure with $X = 2.7$ to get $X = 2.7240302258605$, which we store as B. The area of the region can then be computed by pressing [MATH] [9] to get the integration command, and entering $fnInt(Y3, Y2, X, A, B)$ to get $Area = 16.9461671$. ∎

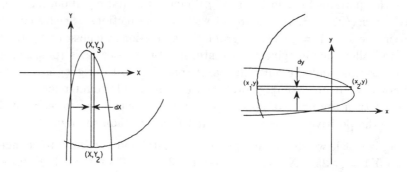

Before we leave this example, let us view what we have done from the actual perspective rather than the reversed horizontal and vertical roles observed on our

5.1. AREA AND VOLUME

calculator. The vertical rectangle on our calculator is actually a horizontal rectangle like the one shown in the right hand figure on the previous page. A "thin" horizontal rectangle always has its right end point on the parabola, and its left end point on the left half of the circle. Its width (the small dimension) is dy and its length is $x_2 - x_1$, where (x_2, y) is a point on the right boundary and (x_1, y) is a point on the left boundary. The area $(x_2 - x_1)\, dy$ of the thin rectangle would have to be expressed in terms of y in order to evaluate the integral $\int_a^b (x_2 - x_1)\, dy$, which would then give the area of the region.

This is how we would have done the problem, if we had managed to graph this region without reversing the horizontal and vertical. Our first impulse might have been to try to sweep out the region with thin vertical rectangles, but notice that it would be very difficult to use vertical rectangles in the actual plot (the right hand figure). The formula for the length of a "thin" rectangle located at a point x on the x-axis appears to change two times (actually three times) depending on what curves the top and bottom sides of the rectangle intersect. (The third change is quite small.)

Example 5.2 *Find the volume of the solid of revolution formed by revolving about the line $x = 10$, the region bounded by the curves $y = x^2 - 8x + 26$, and $y = 4 + 12x - x^2$*

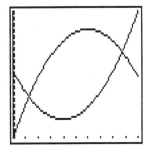

Enter $f(x) = x^2 - 8x + 26$ as $Y1$, and $g(x) = 4 + 12x - x^2$ as $Y2$ in the Y= menu. We adjust the graphing window, eventually choosing $X_{min} = 0$ and $X_{max} = 10$ in the WINDOW menu, to get the screen shown under ZoomFit.

A "thin" vertical rectangle located at the point x on the x-axis inside this region and revolved about the line $x = 10$ would generate a shell. Using the graph screen, it helps to make a pencil and paper sketch, similar to the one shown below, which contains the region and this "thin" vertical rectangle. Collecting shells with respect to x, means integrating with respect to x, and so all of the variables must be expressed in terms of x. The volume of a shell is $v = 2\pi RLT$, where R is the radius, L is the length of the shell, and T is its thickness.

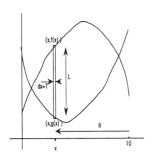

The thickness, $T = dx$, is an assumed part of the integration command, and so it will not be a part of our expression for v. Collecting and adding up, in the sense of integration, all of the volumes of the "thin" shells means integrating $v = 2\pi RL$ once we have R and L in terms of x.

The vertical length L is $L = Y2 - Y1$ (the y-coordinate above minus the y-coordinate below), and the radius R is the horizontal length $R = 10 - x$ (the right hand x-coordinate, where the axis of rotation is located, minus the left hand x-coordinate, where the rectangle is located). Before we can compute the integral, we must determine the x-coordinates of the intersection points of the two curves. To do this, we return to the graph screen, press $\boxed{\text{TRACE}}$ and move the trace cursor to the points of intersection to get $X \approx 1.2$ and $X \approx 8.7$. The $SOLVER$ menu can then be used to get more accurate values. From the SOLVER menu screen, enter $eqn : 0 = Y2 - Y1$, enter $X = 1.2$, leave the cursor on this line and press $\boxed{\text{ALPHA}}$ ($\boxed{\text{SOLVE}}$) to get $X = 1.2583426132261$. This value should be stored as A so go immediately to the home screen and press $\boxed{\text{X,T,}\theta\text{,}n}$ $\boxed{\text{STO}\blacktriangleright}$ $\boxed{\text{ALPHA}}$ ($\boxed{\text{A}}$) $\boxed{\text{ENTER}}$. Return to the SOLVER menu and repeat this procedure to get the second intersection point $X = 8.7416573867737$, which we store as B from the home screen. The volume can then be computed by entering $fnInt(2\pi(10-X)(Y2-Y1), X, A, B)$ to get a volume of $vol = 4388.444987$. ∎

5.2 Mass and Center of Mass

Example 5.3 *A swimming pool has an elliptical shape as shown in the picture. The water is 3 feet deep at one end, 12 feet deep at the other end, and the depth changes linearly from one end to the other. How much water (in cubic feet) is in the pool, and what is its weight in pounds?*

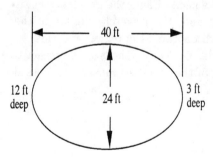

The equation of an ellipse centered at the origin with a horizontal major axis of length 40 and a vertical minor axis of length 24 is

$$\frac{x^2}{20^2} + \frac{y^2}{12^2} = 1$$

It is easy to see that the ellipse crosses the x-axis at $(\pm 20, 0)$, and crosses the y-axis at $(0, \pm 12)$. Using cuts of thickness dx, perpendicular to the x-axis, the

5.2. MASS AND CENTER OF MASS

(3-dimensional) pool can be partitioned into thin, essentially rectangular boxes, or "slabs" of water. The volume of this rectangular "slab" of water is easily expressed as the product of its three dimensions.

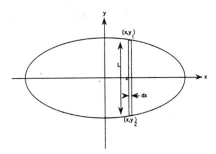

The three dimensions of the slab are L, dx, and the dimension, D (not visible in the diagram), which is perpendicular to the plane of the paper, and represents the water depth of the pool at x. The dx is an assumed part of the integration command, and so the volume of the thin slab takes the form $v = LD$. We add (in the sense of integration) the volumes of all of the thin slabs to form the total volume of the water. The answer, naturally, is expressed in cubic feet, and the total weight of the water is obtained by multiplying its volume by water's density of 62.5 pounds per cubic foot.

Both L, and D must be expressed in terms of x in order to do the integration. Solving the equation of the ellipse for y in terms of x, gives us $y = \pm\frac{3}{5}\sqrt{20^2 - x^2}$, so that

$$L = y_1 - y_2 = y_1 + y_1 = 2y_1 = \frac{6}{5}\sqrt{20^2 - x^2}.$$

The relationship between D and x is linear. We simply write the equation of the line through the points $(x, D), (20, 3), (-20, 12)$. This gives

$$D = 3 + \frac{12 - 3}{-20 - 20}(x - 20) = 3 - \frac{9}{40}(x - 20).$$

If we entered L and D as functions in the Y= menu, it would be easier to enter the integral we wish to compute, but this is not a necessary step, nor does it save keyboard activity. Consequently, we compute the volume of water directly (notice how L and D appear in the integral) by entering

$$fnInt(6/5 * \sqrt{(20 \wedge 2 - X \wedge 2)} * (3 - 9/40 * (X - 20)), X, -20, 20),$$

to get $vol = 5654.866777 ft^3$. The weight of the water, therefore, is $W = 62.5 \times vol = 353429.1736 lbs$ (first press [2ND] [ANS] to return the value of vol to the active line on the home screen). ■

Captain Ralph is an outer-space fighter pilot, with whom we will share an occasional adventure in the future. Our first encounter with him is hardly an adventure, but it would be if he fails in his mission.

Example 5.4 *Captain Ralph is piloting a small shuttle between a space station orbiting Mars and the planet's surface. The shuttle (standing vertically up) has the shape of the solid of revolution formed by revolving about the y-axis the region in the first quadrant bounded by the coordinate axes and $y = 36 - (9x^4)/4$ (The units are all in meters.) As you can see, the ship is quite narrow, being 36 meters tall and only 4 meters wide at the base. (If you wish to get a "true" picture of the ship by plotting this curve, you should plot with the same scale on both axes and use ZSquare from the ZOOM menu.) The ship is also quite nose heavy, with a mass density of $\delta = 219 + 50\sqrt{y}$ kilograms per cubic meter at level y. (Think of density as the mass of a "thin horizontal slice" at level y divided by the volume of the slice.)*

Captain Ralph has been ordered to transport a 2,000 pound laser gun down to the planet's surface, and Ralph has decided that the only safe way to do this is to place the gun as close to the ship's center of mass as possible. Please help Captain Ralph avoid an adventure by finding this point.

To help Captain Ralph, we must find the total mass M of the ship, and its moment M_x with respect to the x-axis. Then the y-coordinate of the center of mass would be $\bar{y} = M_x/M$. Since the center of mass is on the y-axis, $\bar{y}$ is all we need. To generate these terms, we partition the ship into "thin" horizontal disk shaped slices. We determine the mass m of a slice, and the moment m_x with respect to the x-axis of a slice. The total mass M is then the sum (in the sense of integration) of all the "little" masses m and the total moment M_x is the sum (in the sense of integration) of all the "little" moments m_x.

Enter $f(x) = 36 - (9x^4)/4$ as $Y1$ in the Y= menu. The graph of $f(x)$ on the interval $0 \leq x \leq 2$ offers little insight, but it does lead to a pencil and paper picture, shown below, which may help somewhat (no attention was paid to scale).

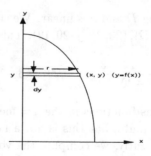

The thickness of a slice is dy, which is omitted from all of our calculations for the same reason as in previous examples. Except for the omission of dy, we have that the volume v of a slice is $v = \pi r^2$, its mass is $m = v\delta$, and its moment with respect to the x-axis is $m_x = ym$ (y is the slice's distance to the axis).

Looking at the above picture, the value of r (a horizontal distance) is $r = x - 0$, which we must express in terms of y. Using analytic techniques, we solve the equation $y = 6 - (9x^4)/4$ for x in terms of y to get

$$r = x = \sqrt[4]{\frac{4}{9}(36-y)} = \sqrt{\frac{2}{3}}(36-y)^{1/4}.$$

5.2. MASS AND CENTER OF MASS

The formulas for $v = \pi r^2$, $m = v\delta$ and $m_x = my$ in terms of y now follow readily. We could express these formulas explicitly in terms of y and then compute the total mass M, and the total moment M_x of the ship directly as integrals with respect to the variable y. The total mass M, for example, could be entered directly in the form $fnInt(m, Y, 0, 36)$, with m expressed in terms of Y. (★86★ There is a better way to do the computations. See the extended comments below. ★86★) Given the complexity of m, this would require a great deal of keyboard activity, and it would leave us with another round of keyboard activity to compute M_x. (If this approach was used, after computing M we could press [2ND] [ENTRY] **to bring the integral for M back to the active line where it could be edited to produce M_x.**)

We choose another approach by entering formulas for r and δ into the Y= menu. To do this, we must use the variable X instead of y. (The current value of $Y1$ no longer plays any role, but we shall leave it in the Y= menu.) Enter $Y2 = \sqrt{(2/3)} * (36 - X) \wedge (1/4)$, and $Y3 = 219 + 50 * \sqrt{(X)}$, so that $Y2$ plays the role of r and $Y3$ plays the role of δ. It follows that $v = \pi r^2 = \pi Y2 \wedge 2$, $m = v\delta = \pi \times Y2\wedge 2 * Y3$, and $m_x = ym = X * \pi Y2 \wedge 2 * Y3$. It is now easy to compute our integrals. The total mass M is computed by entering $fnInt(\pi Y2 \wedge 2 * Y3, X, 0, 36)$ to get $M = 119344.7077$. Press [2ND] [ANS] [STO▸] [ALPHA] [M] to store this value to M. The total moment M_x with respect to the x-axis is computed by entering $fnInt(X * \pi \times Y2 \wedge 2 * Y3, X, 0, 36)$. **The easiest way to enter this integral is to press the combination** [2nd] [ENTRY] **repeatedly until the integral for M reappears on the active input line.** Then move the cursor to the point over π, press [2ND] [INS] [X,T,θ,n] [×] and we're ready to press [ENTER] to get $M_x = 1910428.901$. Press [2ND] [ANS] [STO▸] [ALPHA] [N] to store this value to N. The y-coordinate of the center of mass is then computed by entering N/M to get $\bar{y} = 16.00765495$. ∎

★★★★86★★★★

One of the advantages of this device over the TI-83 is its ability to **store expressions, rather than just numbers as names.** In addition, **the names can be more than just one letter**, and capital or lower case letters can be used. We can enter the expression for mass, and store it to the letter M or to the name "mass". This feature allows us to express ourselves in a much more natural way. We are, for example, not required to use the letter X to represent a variable; we can instead use the letter y, which is certainly a more natural variable in this problem.

We enter, from the home screen, the above formulas for r and δ exactly as they are expressed above in terms of the variable y (we use the letter d instead of δ.) We want to generate lower case letters, so press [2ND] [alpha] [R] [ALPHA] [=] to enter $r =$ onto the active line of the home screen. Follow this by entering the expression for r in a straight forward way, pressing [2ND] [alpha] [Y] to enter the variable y in the appropriate place. When the formula is entered, press [ENTER] and our **calculator responds with the word "done"**. Press [2ND] [alpha] [D] [ALPHA] [=] to enter $d =$ and repeat this procedure to enter the formula for $d = \delta$. The rest is now quite striking in its simplicity.

We could use m for mass, but just to show the features of the calculator, we use the name "mass" instead. To enter the formula $mass = \pi r^2 d$, press [2ND] [alpha]

68 CHAPTER 5. APPLICATIONS OF INTEGRATION

[alpha] (pressing [alpha] twice locks the alpha mode), and then (M) (A) (S) (S) (=) [alpha] (to release the alpha lock). This places *mass =* on the active line. Follow this with the rest of the formula by pressing [2ND] [π] [2ND] [alpha] (R) [^] [2] [×] [2ND] [alpha] (D) [ENTER]. Our calculator responds again with the word "done". Enter the formula *moment = y × mass* in a similar way.

The total mass is the sum, in the sense of integration, of all the "little masses". To find total mass press [2ND] [CALC] [F5] to get the *fnInt* command, and then enter $fnInt(mass, y, 0, 36)$. This gives us the mass we computed above. To find total moment, we add, in the sense of integration, all of the "little moments" by entering $fnInt(moment, y, 0, 36)$, and this gives us the total moment we computed above. These numbers can be stored using the [STO▶] key, and the problem can be finished in the same way it was finished above. ∎

★★★★ 86 ★★★★

5.3 Miscellaneous Problems

Example 5.5 *A water tank, filled with water, has the shape of a circular cylinder with a radius of 20 feet and a length of 80 feet. It lies on its cylindrical side, so that the flat circular ends are vertical. Determine the total force acting on one of the flat circular ends.*

If a flat plate of area $A\ ft^2$ is subjected to a constant pressure of $p\ ^{lbs}/_{ft^2}$, then the total force acting on the plate is $F = pA$. Water pressure at a depth of $h\ ft$ below the surface is $p = \rho h$, where $\rho = 62.5$ is the density of water in pounds per cubic feet.

We use the region R inside the circle $x^2 + y^2 = 20^2$ as a model for one of the circular ends of the tank. The water depth at a point (x, y) inside of this region is $h = 20 - y$. Water pressure is not constant over the entire region, so the formulas above do not apply, at least not directly. Pressure is, however, constant along horizontal lines of constant depth. This prompts us to partition the region into a large number of very narrow horizontal strips. A typical one is shown in the figure.

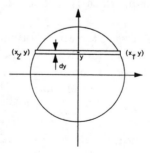

Depth is essentially constant over the entire strip, and so pressure is essentially constant. This means that the simple formula of *force = pressure × area* can be used to approximate the force f acting on the thin strip. This leads to the formula

5.4. EXERCISE SET

$f = 62.5(20-y)(x_1-x_2)\,dy$ for the force acting on the strip. To determine the total force acting on the flat circular wall, we add up in the sense of integration, all of the little forces f. Naturally, the expression $L = x_1 - x_2$ for the length of a strip must be expressed in terms of y before we can integrate with respect to y. Solving the equation of the circle for x in terms of y leads to $L = 2\sqrt{20^2 - y^2}$. To compute the total force acting on the end of the tank we enter

$$fnInt(62.5*(20-Y)*2*\sqrt{(20\wedge 2 - Y\wedge 2)}, Y, -20, 20)$$

to get a total force of 1570796.327 pounds. ∎

So far, all of the applications we have considered were a consequence of the definition of a definite integral as a limit of its Riemann sums. Many other applications are a consequence of interpreting an integral as an antiderivative.

Example 5.6 *A new mutual fund is growing at the rate of* $.83 - 1/\sqrt{(t+5)}$ *in millions of dollars per year, where t is the age of the fund in years. If the initial value of the fund was $94,000, how long will it take for the fund to reach a value of $7,000,000 ?*

If r denotes the above rate, and if v denotes value at time t in millions of dollars, then $r = \frac{dv}{dt}$.

We enter $r = .83 - 1/\sqrt{(t+5)}$ as $Y1$ in the Y= menu, and enter $Y2 = fnInt(Y1, X, 0, X) + .094$. We deactivate $Y1$, since we are not interested in its values. Notice that $Y2$ is an antiderivative of r whose value at time $t = 0$ is 0.094 in millions of dollars (or, in other words $94,000). We wish to find the time X for which $Y2 = 7$ (in millions of dollars). One way to do this is to create a table of values for $Y2$ and look for an X having a $Y2$-value near 7. Initially we guess, by pressing [2ND] [TBLSET], setting $X = 0$, ΔTbl= 10, and then pressing [2ND] [TABLE]. It quickly follows that the time is between $X = 10$ and $x = 20$. To get X a little closer, press [2ND] [TBLSET] again, reset $X = 10$, ΔTbl= 1, and press [2ND] [TABLE] to find that X is between $X = 13$ and $X = 14$. That is close enough first guess. Press [MATH] [0] to get the SOLVER menu, enter $eqn: 0 = Y2 - 7$, $X = 13.5$, and press [ALPHA] [SOLVE] to get $X = 13.217049654443$. It follows that 13.217 years after the initial investment, the fund will be worth $7,000,000. ∎

5.4 Exercise Set

1. Find the area of the region in the first quadrant, bounded by

$$y = \frac{20}{x^2+1}, \quad x = 2, \text{ and } y = 1.$$

Verify your answer by an alternate calculation.

2. Find the area between the curves

$$y = 3x^4 + 4x^3 - 54x^2 - 108x + 1, \text{ and } y = -75 - 36x.$$

3. Find the area between the curves
$$x + 2y^2 - 28y + 96 = 0, \text{ and } x^2 - 8x - 3y + 7 = 0.$$

4. Find the volume of the solid of revolution formed if the region in the first quadrant, bounded by the curves $y = \frac{20}{x^2+1}$, $x = 2$, and $y = 1$ is revolved about
   a) the x-axis   b) the y-axis
   c) the line $x = -4$ d) the line $y = 6$.

5. Verify each of your answers in Problem 4 by an alternate calculation.

6. Find the length of the curve defined by the equation $y = (9 - x^2)\cos(3x)$ on the interval $[-3, 3]$. Try to avoid excessive keyboard activity.

7. Find the volume of the torus generated by revolving about the x-axis the circle of radius 3 centered at the point $(0, 7)$ on the y-axis. Verify your answer with an alternate calculation.

8. Use an integration technique to compute the surface area of a sphere of radius 11.

9. Find the surface area of the torus generated by revolving about the x-axis the circle of radius 3 centered at the point $(0, 7)$ on the y-axis. Verify your answer with an alternate calculation.

10. A pyramid with a horizontal square base and a height of 50 meters, has horizontal cross sections which are squares of side length $25 - h/2$ meters at a height of h meters.
a) Find the volume of the pyramid.
b) If the mass density of the cross section at height h is $\delta = 3.2 + \sqrt{h/3}$ in metric tons per cubic meter, find the mass and center of mass of the pyramid.

11. The region bounded by the curves $y = x^2 - 8x + 8$, and $y = 34x - x^2 - 45$ has a mass density of $\delta = 7 + \sqrt{x}$ (in grams per square centimeter) at each of its points on a vertical line passing through x on the x-axis. Find the mass, and the center of mass of the region.

12. Find the length of a thin wire defined by the curve $y = \sqrt{x} + x\sin(x)$ for x in the interval $[1, 10]$, where x and y are in centimeters. Find its mass and center of mass if the mass density is $\delta = 12 - .73x$ in grams per centimeter at the point on the curve corresponding to x.

13. A flat, circular, glass port hole with a radius of 9 inches has been built into the side of a pool so that underwater activity can be monitored. The top edge of the window is 7 feet below the water surface. What is the total force acting on the glass.

14. A diving bell having a vertical, flat, semicircular, glass window with a radius of 4 feet is used to take tourists down into the ocean to view the wonders of nature. The straight side of the semicircle is the horizontal bottom edge, and the window is built to safely withstand a force of 87,000 pounds. How deep (measure from the water surface to the bottom of the window) can the diving bell safely dive? If the

5.4. EXERCISE SET

diving bell is deep enough, one would expect, intuitively, that pressure is almost constant across the entire window, in which case we could get the force acting on the window simply by multiplying pressure times area. Discuss this matter. Is this a reasonable approximation, at the maximum safe depth?

15. Executives for a logging company are considering the construction of a railroad line from a point where raw lumber is collected to its processing mill, and a map of the region involved is being studied to decide where to build the rail line. A rectangular coordinate system has been placed on the map with the origin at the center of the region. The units are in miles. The raw lumber is collected at the point (0,10), and the processing mill is at the point (20,0). Management must choose between two possible routes labeled C_1, and C_2 below which connect these two points. For each of the two rail lines, company engineers have determined a formula for the approximate force that would have to be exerted at the point (x, y) on the road in order to pull a "standard" train at a constant 30 mph pace throughout the course. These approximate force formulas are labeled f_1, and f_2 below. This information will be used to determine the total work energy required to move a standard train over each of the two rail lines under consideration. Management has decided to choose the rail line, which requires less total energy. The road C_1, and the force f_1 (in thousands of pounds) at the point (x, y) on this road are defined by

$$C_1 = \{(x,y) | y = \frac{20 - x}{2x + 2}, 0 \leq x \leq 20\}, f_1 = 317 - x^2 + 10x - 0.4y^2 + 0.8y,$$

The Road C_2, and the force f_2 (in thousands of pounds) at the point (x, y) on this road are defined by

$$C_2 = \{(x,y) | y = \frac{2000 - x^3 + 30x^2 - 300x}{200}, 0 \leq x \leq 20\},$$

$$f_2 = 132 - 2x^2 + 40x - 3y^2 + 30y.$$

Notice that both curves begin and end at the points (0,10) and (20,0). It is now your job to supply management with these total energy numbers.

16. A model rocket is fired straight up from the ground. The rocket's engine fires for 10 seconds, producing an acceleration $a(t) = 100 + 58\sqrt[3]{t-5}, (0 \leq t \leq 10)$. At $t = 10$ seconds, the engine turns off and the motion of the rocket is subject to just gravitational acceleration. How high does the rocket go, and how long does it take to reach the high point? What is its maximum speed? How long does it take to fall back to earth, and at what velocity does it strike the earth? Would this be a good toy for a 10 year old child? We cannot use our calculator to compute $fnInt(Y2, X, a, b)$ (a and b given constants), if $Y2$ is a an antiderivative of $Y1$ expressed in the form $Y2 = fnInt(Y1, X, a, X)$. On this problem, analytical methods must be used to compute at least one antiderivative exactly.

17. A mineral survey of private farm land, suggests that the land contains approximately 500,000 barrels of oil. The owner of the land plans to install a low capacity oil well, and the survey crew has estimated that it would be capable of producing oil

at the rate of $(\frac{813}{\sqrt{t+7420}} - 1)$ barrels per hour, where t is the age of the well, in hours. The machinery will require very little attention to operate, and so it will be allowed to continue pumping until the production rate drops below 1 barrel per hour, at which time the well will be abandoned. How long, in years, will the well produce oil? How much oil will be recovered, and how much will remain underground when the well is abandoned?

Chapter 6

The Transcendental Functions

According to the dictionary, the word transcendental means something that is beyond common thought or experience, something that is mystical or supernatural. In mathematics, a transcendental expression is an expression which is not algebraic. It is an expression which is somehow beyond algebra. An expression $f(x)$ is algebraic if $y = f(x)$ satisfies an equation of the form

$$a_n(x)y^n + a_{n-1}(x)y^{n-1} + \ldots a_2(x)y^2 + a_1(x)y + a_0(x) = 0,$$

where the expressions $a_0(x), a_1(x), \ldots, a_n(x)$ are all polynomials in x.

The trigonometric functions are all transcendental (beyond algebra), although this is not entirely obvious. In addition to the trigonometric functions, there are several other special transcendental functions which are widely used in mathematics. They will be introduced in this chapter, and their properties relative to calculus will be studied.

Several of these special functions are inverses of other special functions, and so we begin with a section on inverse functions. The chapter ends with a study of the long anticipated *L'Hopital's Rule*. This important rule gives us "the final word" on how to evaluate limits of expressions, when the limits are not obvious. *L'Hopital's Rule* could have been discussed much earlier in the development of calculus, but it is placed in this chapter because so many really mysterious limit examples involve transcendental functions.

6.1 Inverse Functions

Finding the inverse of a function f involves nothing more than solving the equation $y = f(x)$ for x in terms of y. Ideally, it would be nice to have an analytic formula for the inverse, but our calculator in incapable of performing symbolic manipulation. Actually, our ability to solve equations is severely limited by mathematics itself. While mathematicians speak confidently about the existence of solutions, actually finding solutions is another matter. Frequently, solutions cannot be found, even though they clearly exist, and this happens to be the destiny of mathematics, and not a reflection of the person doing the work. As we mentioned on page 36, even the solution to an equation as straightforward as a polynomial may be impossible to find.

Example 6.1 *Find the intervals where*

$$f(x) = \frac{20x}{x^2 - 2x + 4}$$

is one-to-one, and identify three separate branches of the inverse function.
a) Evaluate, if possible, each branch of the inverse function $f^{-1}(x)$ at $x = \pm\sqrt{7}$.
b) Two branches of the inverse have domains which include the interval $[1, 3]$. For each of them, create a table of values of $f^{-1}(x)$ for x in the set $D = \{1, 1.1, 1.2, \ldots, 3\}$.
c) Verify that the identity $f(f^{-1}(x)) = x$, holds for each x in the set D defined in part (b).
d) Verify that for the appropriately chosen inverse, the identity $f^{-1}(f(x)) = x$ holds for each x in the set $R = \{30, 32, 34, \ldots 80\}$.

A great deal of insight can be gained by first turning our attention to a graph of $f(x)$, so we enter $f(x)$ as $Y1$ and graph it under ZoomFit with $X_{min} = -10$ and $X_{max} = 10$ to get the screen shown.

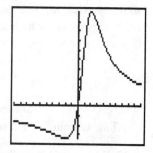

This plot supplies a wealth of information. The function f appears to be one-to-one on three separate intervals, and so there should be three separate inverses. How do we know, however, that this plot does not have hidden features too small to see (or features beyond this window) which would make these conclusions false? To be on the safe side we first look at the sign of $f'(x)$. The derivative is easy enough to compute analytically, and from the formula

$$f'(x) = \frac{4 - x^2}{(x^2 - 2x + 4)^2},$$

it is clear that f is a one-to-one increasing function on the interval $[-2, 2]$, because fp is positive on this interval, and f is a one-to-one decreasing function on the intervals $(-\infty, -2]$ and $[2, \infty)$, because fp is negative there.

The intervals for the inverses will be determined by the high and low points on the graph. The derivative gives us an easy means to select these points, which we denote as $a = -2$, $b = 2$, $\alpha = f(a) = -20/6$, $\beta = f(b) = 10$.

From the graph of $f(x)$, we can see the domains and ranges of all three of the inverses, and so we let g_1, g_2, and g_3 be the inverses to f on these various intervals with inverses and intervals paired as follows.

$$(-\infty, a] \;\underset{g_1}{\overset{f}{\rightleftarrows}}\; [\alpha, 0) \qquad [a, b] \;\underset{g_2}{\overset{f}{\rightleftarrows}}\; [\alpha, \beta] \qquad [b, \infty) \;\underset{g_3}{\overset{f}{\rightleftarrows}}\; (0, \beta]$$

In this particular example, formulas for the inverses could be determined, but only with a fair amount of analytic work. In another example, formulas might just

6.1. INVERSE FUNCTIONS

as well be impossible to find. Regardless of the situation, it is a straight forward process to evaluate an inverse at a given number. From the graph and diagram above, it follows that $x = \sqrt{7}$ is in the domain of g_2 and g_3, while $x = -\sqrt{7}$ is in the domain of g_1 and g_2.

To evaluate $g_2(\sqrt{7})$, we use the SOLVER menu to find the unique solution to the equation $f(x) = \sqrt{7}$ for x in the interval $[-2, 2]$. There are, of course, two solutions to this equation, but as long as we enter an appropriate first guess, our solution will be in the interval $[-2, 2]$. There is no need to be careful. We could choose any point in the interval as a first guess, although it is probably wise to stay away from the end points of this interval. Press [MATH] [0] to bring up the SOLVER screen, enter $Y1 - \sqrt(7)$ at $eq: 0 =$, enter 0 at $X =$ (or some other value in the interval $[-2, 2]$) and press [ALPHA] ([SOLVE]) to get $g_2(\sqrt{7}) = 0.43856148441513$. In much the same way we get $g_2(\sqrt{7}) = -0.84924973799098$. The process of evaluating $g_1(-\sqrt{7})$ is similar, except that we must change the start up (first guess) value of X to a point in the interval $(\infty, -2)$. Using a start up value of $X = -10$ quickly leads to $g_1(-\sqrt{7}) = -4.7100397221934$. In a similar way, using a start up value of $X = 10$, we determine that $g_3(\sqrt{7}) = 9.1207279757716$.

Creating a table of values for $g_2(x)$ and $g_3(x)$ for $x = 1, 1.1, \ldots, 3$, involves much the same kind of work. We simply solve the equation $f(x) = y$ for the range of values $y = 1, 1.1, \ldots, 3$.

The most convenient way to do this is to create a program. The program is given below, and it can, if desired, be loaded onto your calculator, and used with little attention paid to the structure of the program itself. This is a good opportunity, however, to learn how to write programs on your calculator. The TI-83 Guidebook is an excellent reference to learn the various programming commands, and it should be consulted. Lists are used extensively in the following program. We began to use them on our calculator on page 47. Return to that page or consult the TI-83 Guidebook, if you wish to review the process of entering and using lists on your calculator.

★★★★ 86 ★★★★

It is notably easier to write programs with the TI-85,86. The menu commands are much more accessible. In particular, lists are easier to create and use, because the prefix L does not have to be attached to the names of lists as it does on the TI-83. Consult the Chapters on programming and lists in the TI-86 Guidebook

★★★★ 86 ★★★★

Press [PRGM] ▶ ▶ [ENTER] and enter a name for the program. The calculator will be in locked alpha mode during this step. After we name our program "INVLIST", we press [ENTER] and the cursor moves to a new line containing a colon. This is where the first line of the program will be entered. After each line is entered, press [ENTER], to move the cursor to a new programming line.

To enter the commands, place the cursor where the command should be entered on the program line, press [PRGM], select either the CTL or I/O submenus, select the appropriate item in the submenu, and press [ENTER] to enter that command on the programming line.

Some commands are available in other menus, most notably, the LIST menu, and they can be retrieved in the usual way be pressing the appropriate menu key. Other commands are available only in the CATALOG menu. It is a vast menu containing all of the commands on the calculator. Press [2ND] [[CATALOG]] and the calculator will automatically be in alpha mode as it opens the CATALOG menu. Follow this by pressing, for example, ([s]) to move to the commands beginning with the letter s.

The following diagram shows the placement of the various sets (lists) used in the program relative to the function f and a branch $g_k(x)$ of $f^{-1}(x)$.

$$W \subset [A, B] \underset{g_2}{\overset{f}{\underset{\leftarrow}{\rightarrow}}} [\alpha, \beta] \supset V$$

In the applications of the program, it may help to draw similar diagrams in order to keep track of the various sets involved.

Entering a program requires a great deal of keyboard activity. Mistakes are likely to be made, especially if you are creating your own program. We will discuss this matter in a moment, but for now, let us enter the program.

PROGRAM: INVLIST

:Disp "A" — This displays the letter A on the screen before input.
:Input A
:Disp "B"
:Input B — The letters A and B represent an interval $[A, B]$ which contains all of the values $f^{-1}(x)$ under consideration.

:Disp "LIST"
:Input V — The letter V represents the list of points x in the table of values for $f^{-1}(x)$.
:dim(LV)$\to N$ — The letter N represents the number of entries in the list LV.
:DelVar LW — This clears the value of LW from memory.
:$1 \to J$ — This sets the index for our program loop to 1.
:Lbl Q — The Lbl command marks the start of a program loop, and the letter Q becomes the name of the loop.
:solve($Y1-$L$V(J), X, \{A, B\}$) — This solves the equation $Y1-$L$V(J)$ for X in the interval $[A, B]$. The notation L$V(J)$ represents the Jth entry in the list LV. Recall that $Y1 = f(X)$.
:Ans$\to$L$W(J)$ — This stores the solution as the Jth entry in the list LW of values of the inverse.
:IS$>(J, N)$ — Note that the notation IS$>$ is all part of one command in the CTL submenu of the PRGM menu. To begin with, J is increased by 1. Then the command asks, "Is $J > N$?" If the answer is "yes", then the next line is skipped. Otherwise the program continues to the next line.

6.1. INVERSE FUNCTIONS

:Goto Q This returns the program to the beginning of the program loop named Q.

:List▶matr($_LV$,$_LW$,$[I]$) This turns our answer into a matrix called $[I]$. We will discuss this step below. Press MTRX and enter the name $[I]$ (I for inverse) through the NAMES submenu. (★86★ You are free to use more natural names for matrices on this device. ★86★)

:disp "MATRX [I]" This reminds us where to look for the output.

The second from last step in the program is included only for display purposes. Matrices are important mathematical objects, but they are seldom studied in calculus. On the other hand, the output of a matrix looks like a table, and so it is a convenient way to display the output. This program step creates a table with a left hand column of $_LV$ and a right hand column of $_LW$. After we get our table of values, we will discuss how to view it.

We could have skipped this step, and left the output in the form of a list. The STAT EDIT screen would be another excellent way to display the output in a very readable form as a table. We used this approach page 48. A matrix approach was used instead only because it provided us with a good opportunity to introduce a few new features on our calculator.

After this program is entered, press 2ND [QUIT] to return to the home screen where we can now create our table. The list of points $D = \{x = 1, 1.1, \ldots, 3\}$, which are to appear in our table, are only in the domains of $g_2(x)$ and $g3(x)$. We begin with $g_2(x)$, which takes its values in the set $[A, B] = [-2, 2]$.

Before we can run this program, $Y1 = f(x)$ must be entered in the Y= menu. We could enter the list D when it is called for in our program, but it may seem easier to do this step ahead of time. Press 2ND [LIST] ▶ 5 to bring up the sequence command. Enter $seq(1 + .1 * J, J, 0, 20)$, and then press STO▶ ALPHA ([D]) ENTER to store the list to D.

From the home screen, press PRGM, select the EXEC submenu, move the cursor to the program named INVLIST, press ENTER to move the name to the home screen, and press ENTER again to activate the program. The letter A appears on the home screen followed by a question mark. Press (−) 2 ENTER. Our calculator sets $A = -2$, and brings up the letter B with a question mark. Pressing 2 ENTER sets $B = 2$, and brings up the word "LIST" followed by a question mark. At this point, we could directly create a list, but we have done this in advance, so we enter $_LD$. (This list becomes the list $_LV$ in the program, but we don't need to know that.) After entering $_LD$, press ENTER, the program runs, and eventually responds by displaying "MATRX [I]" and then "done" on the home screen.

There are several ways to view the table $[I]$, but the best viewing screen is obtained by pressing MTRX ▶ ▶ to select the EDIT submenu, and then scrolling down to select $[I]$. Press ENTER to get the screen shown. We can scroll through this screen in much the same way that we maneuvered through a table in the TABLE menu. Looking at the table we can see, for example, that $g_2(1) = f^{-1}(1) = .1833461736$.

```
MATRIX[I] 21x2
[ 1      .1833       ]
[ 1.1    .20018      ]
[ 1.2    .2168       ]
[ 1.3    .23322      ]
[ 1.4    .24943      ]
[ 1.5    .26547      ]
[ 1.6    .28132      ]↓
1,2=.1833461736...
```

In a perfect world, this is what is suppose to happen, but if there is a mistake in our program, it may well crash when we try to run it. If it does, select the "goto" option, and the cursor will move to the point in the program where the mistake was encountered. This helps a great deal in the repair process. Correcting the mistakes and fine tuning the program will almost always involve moving to a PROGRAM EDIT screen without the aid of a "goto" option. Press PRGM ▶ to enter the EDIT submenu, select the program and press ENTER to move into the program where it can be edited.

Before we turn our attention to a table of values for $g_3(x)$, we take the opportunity to do *part (c)* of this example. Our notation is already set up to easily show for the branch $g_2(x)$ of the inverse $f^{-1}(x)$ that $f(g_2(x)) = x$ for each x in the set $D = \{1, 1.1, \ldots, 3\}$. The set D is already entered on our calculator as the list LD, and the list LW is the corresponding list of values $g_2(x)$ as x ranges over LD. (Incidentally, we would know this about LW only by reading the program.) The point $x = 1.7$, for example, is $x = 1.7 =$L$D(8)$), the 8*th* element in LD. The corresponding point $g_2(1.7) =$L$W(8)$) is the 8*th* element in LW. From the home screen we enter $Y1($L$W(8))$ and press ENTER to get 1.7. **More impressive, recall that functions on our calculator are listable.** All we have to do is show that the list $f($L$W) =$LD. Simply enter $Y1($L$W)$ on the home screen and press ENTER to get the output $\{1, 1.1, 1.2, \ldots, 3\}$ which is the list LD.

As an interesting addition, we mentioned above that we used the list LW only because we read the program. What would we do if we didn't read the program? The list of values $g_2($L$D)$ is still accessible to us, even if we don't know that this is the list LW. By looking at the output $[I]$ it is clear that this list is the second column of $[I]$. To access this column, from the home screen, and set it equal to LW (or any other list name), press MTRX ▶ to access the MATH submenu, press 8 to enter the command Matr▶list on the home screen. Then enter Matr▶list($[I], 2, C$) and press ENTER. We used the letter C instead of W just to demonstrate that it is unnecessary to know about the letter W created in the program. **The command takes column 2 from $[I]$ and stores it to the list LC.** To finish *part (c)* in the example, we could have entered $Y1($L$C)$ instead, and pressed ENTER to see that the output returned the list LD.

Getting a table of values for $g_3(x)$ instead of $g_2(x)$ for x in $D = \{1, 1.1, \ldots 3\}$, involves the same methods we used above. A reasonable choice for A would be $A = 2$, but what should we use for B? Press GRAPH and look at the graph of $Y1 = f(x)$. Press TRACE and move the trace cursor to the right until $Y1$ falls below 1. This happens near $X = 23$, so we set $B = 24$. The remaining details are omitted.

Finally, to show *part (d)* of the example, enter the list $seq(30 + 2*J, J, 0, 25)$, press STO▶ ALPHA (R) ENTER, to store the list as R. Enter $Y1(\text{L}R)$, and press STO▶ ALPHA (D) ENTER to store the list $f(R)$ as D. We can now use the program to create a list $g_3(\text{L}D)$ of inverse values for points in D. Can you see why g_3 is the appropriate branch of the inverse function to use? What should we use for A and B? Recall that the interval $[A, B]$ selects a branch of the inverse function. It identifies an interval $[A, B]$ over which $f(x)$ is one-to-one. Clearly, in the present case, we must chose $[A, B]$ to at least include the interval $[30, 80]$. Press PRGM, select the INVLIST program, press ENTER ENTER, and at the prompts, enter 20 for A, 90 for B, and LD for "LIST". When the program is finished, press MTRX ▶ ▶, scroll down to $[I]$ and press ENTER. The table of values of $f^{-1}(f(x))$ for points x in the set LR will be the second column in this table. Notice that it does, indeed, return R. ∎

As a final topic in this section on inverse functions, we draw your attention to a theorem in your calculus text, which allows one to evaluate the derivative of an inverse function at a point, even if a formula for the inverse cannot be found. If $g = f^{-1}$ and if $b = f(a)$, then, according to this result,

$$g'(b) = \frac{1}{f'(a)}.$$

It is usually a straightforward matter to show that an inverse exists, but it is frequently difficult or impossible to actually find a formula for the inverse. This result, then, can be very useful.

Example 6.2 *Show that the function $f(x) = 2x + \sin(x)$ is one-to-one on the whole real line. Compute, approximately, the derivative of $f^{-1}(x)$ at the point $x = 17$.*

The function $f(x) = 2x + \sin(x)$ has $f'(x) = 2 + \cos(x)$ as its derivative. Since $|\cos(x)| \le 1$, it follows that $f'(x) > 0$, so that $f(x)$ is strictly increasing on the interval $(-\infty, \infty)$. This means that $g(x) = f^{-1}(x)$ exists. According to the above theorem, $g'(17) = \frac{1}{f'(a)}$, where $f(a) = 17$. Finding a value for a has become a routine problem. After entering $Y1 = f(x)$, $Y2 = f'(x)$, and deactivating $Y2$, we graph $Y1 = f(x)$, press TRACE and move the trace cursor close to $Y1 = 17$. The corresponding value $X \approx a$ serves as start up value for solving the equation. (There is no need for any accuracy in this step.) Press MATH 0 enter $eq: 0 = y1 - 17$, press ▼ (X will automatically have the X-value of the trace cursor, press ALPHA (SOLVE) to get $X = a = 8.0057471491172$. From the home screen, enter $1/Y2$ (recall that X now has the value of our solution to $f(X) = 17$) to get $g'(17) = .5408865878$.

6.2 Logarithmic and Exponential Functions

It is an interesting oddity of mathematics, that there are elementary and very obvious antiderivatives for all of the power functions $f(x) = x^n$ except in the one case $n = -1$. How strange it is, that we cannot find a simple algebraic function whose derivative is $f(x) = 1/x$. Fortunately, the *Fundamental Theorem of Calculus* gives us a fail safe method to specify the antiderivative of any continuous function, whether or not an analytic formula can be determined. We used this theorem and our calculators extensively in Section 4.2 to specify antiderivatives numerically. Using the same technique now to specify an antiderivative of $1/x$ $(x > 0)$, we define the function $\ln(x)$ by

$$\ln(x) = \int_1^x \frac{1}{t} dt.$$

This function is of more than passing interest, and so it is natural to study its properties. Naturally, the function $\ln(x)$ is available on the keyboard of our calculator, but if we simply use the [LN] key to get its values, we will miss an important opportunity to take another look at the significant role played by the *Fundamental Theorem of Calculus* in solving this important problem of finding an antiderivative for $(1/x)$.

For this reason, let us begin our study by entering $\ln(x)$ in the Y= menu of our calculator in the form

$$Y1 = fnInt(1/T, T, 1, X)$$

We could have used the letter X instead of T, but T is more appropriate for conceptual purposes. It is worth remembering that our calculator does not evaluate this integral (or any other integral) by looking for an antiderivative. Given a value for X it simply uses numerical techniques, like those studied in Chapter 4 to approximate the area under the graph of $Y = 1/T$ from $T = 1$ to $T = X$. It is clear from the definition that $Y1 = \ln(X)$ is defined only for $X > 0$, because the integrand $1/T$ is discontinuous at $T = 0$.

Example 6.3 *a) Use the definition $Y1 = fnInt(1/T, T, 1, X)$ of $\ln(X)$ to evaluate this function at the points $X = 10^{-1}, 10^{-7}, 10^{-8}, 10^{-9}$, close to $X = 0$. Explain why the values are negative.*
b) Evaluate $Y1(X)$ at the large numbers $X = 10, 10^7, 10^8, 10^9$.
c) Compare the values obtained in (a) and (b).
d) Show that $Y1(ab) = Y1(a) + Y1(b)$ for $a = 45, b = 87$.
e) Find a point X such that $Y1(X) = 1$.

In the usual way, press [VARS] ▶ [1] [1] to enter $Y1$ onto the home screen.

$$Y1(10 \wedge -1) = -2.3025855093 \quad Y1(10) = 2.3025855093$$
$$Y1(10 \wedge -7) = -16.11809545 \quad Y1(10 \wedge 7) = 16.11809545$$
$$Y1(10 \wedge -8) = -18.42068073 \quad Y1(10 \wedge 8) = 18.42068074$$
$$Y1(10 \wedge -9) = -20.72326021 \quad Y1(10 \wedge 9) = 20.72326584$$

Why are the values $Y1(X)$ negative for $X < 1$? In this case, $fnInt(1/T, T, X, 1)$ represents the area (a positive number) under the graph of $Y = 1/T$ on the interval $X \leq T \leq 1$. Consequently, $Y1(X) = fnInt(1/T, T, 1, X) = -fnInt(1/T, T, X, 1)$ is negative. Notice how the values over the set of small numbers X are just the negatives of the values over the set of large numbers X. (Because of integral approximation errors, the variation in the digits is more substantial when comparing very small and very large numbers X.) Notice how the entries satisfy the property $Y1(10^n) = n*Y1(10)$. This is solid evidence in support of the rule $\ln(a^n) = n\ln(a)$, which is proved in any standard calculus text.

Next, we enter $Y1(45*87)$ on the home screen to get $Y1(45*87) = 8.272570608$. Entering $Y1(45) + Y1(87)$ gives the same value of 8.272570608. This evaluation supports the general rule $\ln(ab) = \ln(a) + \ln(b)$, which holds for positive numbers a and b. This rule is also proved in any standard calculus text.

To solve the equation $Y1(X) = 1$, press [MATH] [0] to select the SOLVER menu. It is then a straight forward process to enter the equation $eqn : 0 = Y1 - 1$. What should we use for X? We could graph $Y1$ to get a start up value, but it is easier to just look at the above values. The number 1 is roughly one half of the value $Y1(10)$. If the above rule holds in general, then $\frac{1}{2}Y1(10) = Y1(10^{1/2}) \approx 1$. Based on this, we use $X = 3 \approx \sqrt{10}$ for a start up value. With the cursor on the line $X = 3$ press [ALPHA] ([SOLVE]) to get

$$X = 2.7182818284588.$$

The number X is displayed prominently, because of its significance. We talk about this matter next, but the last calculation ends our example, and it also ends our use of $Y1 = fnInt(1/T, T, 1, X)$ as a means of evaluating $\ln(x)$. Pressing the [LN] key is a much faster and more convenient way of getting logarithmic values. ∎

Incidentally, $\ln(x)$ is defined only for positive values of x, but **our calculator will evaluate $\ln(x)$ for negative values of x**. It turns out that $\ln(x)$ can be defined as a complex number when $x < 0$. Such values, however, have no place in calculus, and **they should be avoided**.

The evaluations in the last example, and the rules concerning $\ln(x)$, which are proved in standard calculus texts suggest (more than suggest—prove) that this "area" function defined by way of an integral is, in fact, a logarithm function. Its base then would be a matter of considerable interest. For a general logarithm, $x = b$ satisfies the equation $\log_b(x) = 1$, and so it follows that the above value for X must be the base (approximately) for the natural logarithm $\ln(x)$. This is why the last calculation in the previous example was displayed so prominently. The base is universally denoted by e, and its decimal value, correct to as many digits as are displayed above is

$$e = 2.7182818284590.$$

It is interesting to notice how accurate our above estimate turned out to be.

The inverse of $\ln(x)$ is the exponential function $\exp(x) = e^x$. One can play graphically with these two functions to establish a variety of results mentioned in standard calculus texts. For example, to establish the identities

$$\ln(e^x) = x \ (-\infty < x < \infty), \qquad e^{\ln(x)} = x \ (0 < x, \infty),$$

graph the functions $Y1 = \ln(e^X)$, and $Y2 = e^{\ln(X)}$ (one at a time). Chose a large interval with $X_{min} = -100$ and $X_{max} = 100$. In the first case, we see the line $Y = X$

for $-100 \leq X \leq 100$, and for $Y2$ we only get the graph $Y = X$ for $(0 < X \leq 100)$. The function $ln(x)$ is undefined for $x \leq 0$. This is why the graph of $Y2$ is restricted.

We can use our calculator to offer dramatic evidence of an important property concerning the explosive growth of the exponential function e^x (or, for that matter, any other function of the form a^x for $a > 1$).

Example 6.4 *Show graphically that $e^x \to \infty$ faster than x^n as $x \to \infty$ for $n = 20$. (This result actually holds for every positive constant n, regardless of how large.)*

Take a moment to think about how impressive the growth of e^x must be in order to get large faster than $p(x) = x^{20}$. At $x = 10$, $p(10) = 10^{20}$ is already a hundred thousand trillion, and p(x) itself continues to grow at an ever faster rate for $x > 20$.

We could show this result by graphing the pair of functions $y = e^x$ and $y = x^{20}$, but there is a more interesting way to proceed. We graph instead the function $q(x) = e^x/x^{20}$. If $q(x) = \to \infty$ as $x \to \infty$, then the result follows. Clearly, $q(x)$ will be small for a while. At $x = 3$ for example, 3^{20} will clearly overwhelm e^3. As you might expect, $q(x)$ remains small until x is quite large. It is not easy to graph this function. After entering this function as $Y1$ in the Y= menu, notice what the graph looks like under ZoomFit with $X_{min} = 3$ and $X_{max} = 20$. The graph appears to be large at first, but this is an illusion. Press [TRACE] and move the trace cursor to these points and observe the small Y-coordinates. Alternately, press [WINDOW] and notice how small Y_{min} and Y_{max} are.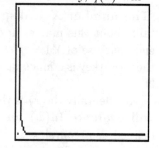

To find out where this graph turns around, we determine where its derivative $q'(x) = 0$. After making important simplifications, we get

$$q'(x) = \frac{e^x(x - 20)}{x^{21}}.$$

This result is quite revealing! The function $q(x)$ decreases for $0 < x < 20$, increases for $x > 20$ and has an absolute minimum at $x = 20$. Based on this result we graph $Y1 = q(X)$ on the interval $20 \leq X \leq 40$. The results are disappointing and are not shown. It turns out that that graph exhibits its increasing property, but its Y-coordinates are exceedingly small throughout the interval.

Rather than use a trial and error approach, we solve the equation $q(x) = 100$ just to see how big x must be to generate this value. Press [MATH] [0] to get the SOLVER menu, and enter $eqn : 0 = Y1 - 100$. Based on the last graph (which is not shown) our start up value for X must clearly be larger than 40. We try a wild guess of $X = 80$, and with the cursor on the line $X = 80$, we press [ALPHA] ([SOLVE]) to get $X = 95$ (along with many additional digits which are of no interest to us).

6.2. LOGARITHMIC AND EXPONENTIAL FUNCTIONS

We set $X_{min} = 40$, $X_{max} = 150$, and graph $q(x)$ under ZoomFit to get the screen shown. It may look disappointing, but it supplies all the evidence we need to make our conclusions. The graph looks small except for its right hand side only because the values at this end overwhelm the rest of the graph. **It will not help to use the trace cursor to evaluate the function at points on the curve.** The Y-values take up too much space to be completely displayed at the bottom of the graph screen. To see how big $Y1 = q(X)$ is at the right hand side of the graph window, press [WINDOW] and look at the sized of Y_{min} and Y_{max}. Instead, we go to the home screen and enter $Y1(150)$ to get $q(150) \approx 4 \times 10^{21}$. ∎

Example 6.5 *Find the x-intercept of the line which is tangent to the graph of* $y = \ln(x^3 + 5)$ *at the point on the curve having a y-coordinate of 10.*

We could **use the DRAW menu to graph the tangent line**, but this would not help us locate the desired x-intercept with any precision. Instead, we find the equation of the tangent line. Enter $f(x) = \ln(x^3 + 5)$ as $Y1$ in the Y= menu. After computing $f'(x) = \frac{3x^2}{x^3+5}$ analytically, we enter $Y2 = f'(X)$ in the Y= menu. Since the line through (x_0, y_0) with slope m has equation $y - y_0 = m(x - x_0)$, we can immediately go to the Y= menu, and enter $Y3 = Y1(10) + Y2(10) * (X - 10)$. We deactivate $Y2$ and graph $f(x)$ and its tangent line under ZoomFit with $X_{min} = -20$ and $X_{max} = 20$ to get the screen shown. Press [TRACE] ▲ and move the trace cursor to where the line crosses the x-axis to get $X \approx -13$. For more accuracy, press [MATH] [0] to bring up the SOLVER menu, enter $eqn := 0 = Y3$, move the cursor to the line $X =$ (X already has our approximate value) and press [ALPHA] ([SOLVE]) to get $X = -13.157688448652$. ∎

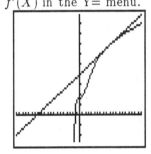

Example 6.6 *Find the area of the region bounded by the graphs of* $y = e^{2x-3}$, $y = 4$, *and* $x = -1$. *Verify with an alternate calculation.*

Remember from our previous example that exponential functions grow large rapidly, so we should pay some attention to the plotting interval. We enter $Y1 = 4$, and $f(x) = e^{2x-3}$ as $Y2$ in the Y= menu. Using the estimate $e \approx 3$, it follows that $e^1 < 4 < e^2$, so we choose $X_{min} = -2$ (a little less than -1) and $X_{max} = 2.5 = 5/2$ (because $f(5/2) = e^2$), and graph the curves under ZoomFit. We could use our imagination to picture a vertical line at $x = -1$, but we could also press, from the graph screen, [2ND] [[DRAW]] [4] (to select the Vertical command). A vertical line appears on the graph screen and the cursor can be used to place it as close as possible to $x = -1$. As we see in the screen shown, this gives a clear picture of the area under consideration.

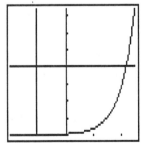

Imagine a thin vertical rectangle located at position X inscribed in the region. Its height (a difference in y-coordinates) is $4 - Y2$, its width (a small difference in x-coordinates) is dx, and so its area is $(4 - Y2)\, dx$. Adding —in the sense of integration—all of the "little" areas, means forming the integral $fn(4-Y2, X, -1, A)$ where A is the X-coordinate of the upper right hand corner of the region. To find A, press [MATH] [0] to select the SOLVER menu, enter $eqn : 0 = Y2 - 4$, set $X = 2$, and press [ALPHA] ([SOLVE]) to get $X = 2.1931471805599$. Without changing X, go to the home screen, and press [X,T,θ,n] [STO▶] [ALPHA] ([A]) to store this value as A. We can now compute the area by entering $fn(4 - Y2, X, -1, A)$ to get $area = 10.7759577$.

An alternate way of calculating this area would be to sweep out the region with thin horizontal rectangles. Imagine a thin horizontal rectangle located at position Y inscribed in the region. Its length (a difference in x-coordinates) would be $X - (-1)$, where X is the X-coordinate of the point $(X, Y2)$ on the curve. Its width (a small difference in y-coordinates) is dy, and so its area is $(X+1)\, dy$. Adding —in the sense of integration—all of the "little" areas, means forming the integral $fn(X+1, Y, B, 4)$ where B is the Y-coordinate of the point at the lower left hand corner of the region. The value of B is easy to get. It is simply $B = Y2(-1)$. Clearly, we must express X in terms of Y in order to evaluate the integral, and this can be done analytically. To solve the equation $y = e^{2x-3}$, for x in terms of y, simply take the logarithm of both sides to get

$$2x - 3 = \ln(e^{2x-3}) = \ln(y),$$

which can then easily be solved for x to get $x = (\ln(y) + 3)/2$. We can now enter the integral using any letter as the variable. It makes conceptual sense to use the letter Y, so this is the letter we shall use. The letter Y should be entered by pressing [ALPHA] ([Y]) and should not be entered by using the VARS menu. The value of $B = Y2(-1)$, on the other hand, must be entered in the usual way by first pressing [VARS] [▶] [1] [2] to enter $Y2$, and then finishing with $Y2(-1)$. We enter $fnInt((\ln(Y) + 3)/2 + 1, Y, Y2(-1), 4)$ to get $area = 10.77595769$. ∎

There is no key on our calculator for evaluating $\log_b(x)$, for $b \neq e, 10$, but this is no cause for alarm. The formula

$$\log_b(x) = \frac{\ln(x)}{\ln(b)},$$

is proved in standard calculus texts, and it can be used effectively to evaluate any logarithm.

Example 6.7 *Solve the inequality* $\log_8(\log_9(x)) \leq \log_9(\log_8(x))$.

To enter a function of the form $L(x) = \log_a(\log_b(x))$ onto our calculators, we must first convert both logarithms to natural logarithms using the above identity. The general properties of logarithms can be used to simplify the result. This gives

$$L(x) = \log_a(\log_b(x)) = \frac{\ln(\log_b(x))}{\ln(a)} = \frac{\ln(\frac{\ln(x)}{\ln(b)})}{\ln(a)} = \frac{\ln(\ln(x)) - \ln(\ln(b))}{\ln(a)}.$$

Using this form with $a = 8, b = 9$, we can now enter $f(x) = \log_8(\log_9(x))$ as $Y1$ and with $a = 9, b = 8$ we can enter $g(x) = \log_9(\log_8(x))$ as $Y2$ in the Y= menu. Both functions are defined for $1 < x < \infty$ (the inner logarithm must have a positive value), and so we can plot both of them under ZoomFit with $X_{min} = 1$ and $X_{max} = 100$. The plot, however is of little use, since the functions are too close together to see which one is smaller. We could zoom-in and eventually decide which one is locally smaller, but we would not be able to generalize our decision beyond a small window, and certainly not beyond X_{max}.

In situations like this, more information can often be obtained by graphing the difference function $H(x) = f(x) - g(x)$ instead. Even though $H(x)$ may be small, one can always scale the window so that it is easy to see whether $H(x)$ is positive or negative. With this in mind, we enter $Y3 = Y1 - Y2$ in the Y= menu, deactivate $Y1$ and $Y2$, and plot $Y3$ under ZoomFit with $X_{min} = 1, X_{max} = 200$. It follows readily that $H(x)$ is negative for $1 < x < 200$.

Are we done? Is $H(x)$ always negative? To answer these questions with any confidence, clearly we must bring a more analytic argument to the table. We begin by determining the slope of $f(x)$ and $g(x)$. Notice that the general double logarithm $L(x)$ displayed above has the relatively simple form $L(x) = c_1 \ln(\ln(x)) + c_2$ where $c_1 = 1/\ln(a)$ and $c_2 = -\ln(\ln(b))/\ln(a)$ are constants. Since

$$L'(x) = c_1 \frac{1}{\ln(x)\, x} = \frac{1}{\ln(a)} \frac{1}{\ln(x)\, x},$$

we can infer that

$$f'(x) = \frac{1}{\ln(8)} \frac{1}{x \ln(x)}, \quad g'(x) = \frac{1}{\ln(9)} \frac{1}{x \ln(x)}.$$

It follows that $g'(x) < f'(x)$ for all $x > 1$. This means that the function $H(x) = f(x) - g(x)$ defined above satisfies $H'(x) > 0$ for all $x > 1$. It follows that the increasing function $H(x)$, which is initially negative, is either always negative, or it crosses the x-axis at a unique point x_0, and satisfies $H(x) < 0$ for $1 < x < x_0$ and $H(x) > 0$ for $x > x_0$. We pull up the SOLVER menu by pressing [MATH] [0]. Enter $eqn : 0 = Y3$, $X = 200$ and press [ALPHA] [SOLVE] to get $x_0 = 334.23922225553$. As a final check, go to the home screen, and without changing the value of X, press [VARS] [▶] [1] [1] [ENTER] to get $Y1 = 0.4677722881$. Enter $Y2$ in the same way to establish that $Y2 = Y1$. It follows that $\log_8(\log_9(x)) < \log_9(\log_8(x))$ for $1 < x < x_0$. The two double logarithms are equal at x_0 and the inequality is reversed on the interval $x > x_0$. ∎

6.3 The Inverse Trigonometric Functions

Only three of the inverse trigonometric functions are available on the keyboard. The remaining three are available from the identities:

$$\sec^{-1}(x) = \cos^{-1}(1/x) \quad \csc^{-1}(x) = \sin^{-1}(1/x) \quad \cot^{-1}(x) = \tan^{-1}(1/x) \quad (6.1)$$

The proofs of these identities, which are problems in the exercise set, follow in a natural way from the idea of an inverse.

The inverse functions are, of course, not globally one-to-one, and so their inverses only exist if their domains are suitably restricted. This complicates the identities

$$f(f^{-1}(x)) = x, f^{-1}(f(x)) = x.$$

If $f(x) = \sin(x)$, for example, the first of these identities holds as long as $f(f^{-1}(x))$ is well defined, namely, for all x in the interval $-1 \leq x \leq 1$.
Enter the function $\sin(\sin^{-1}(x))$ as $Y1$ in the Y= menu, set $X_{min} = -1, X_{max} = 1, Y_{min} = -1, Y_{max} = 1$, and press GRAPH to see the tell tale shape of the curve $Y = X$ that we expect to see.

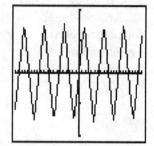

The second identity, on the other hand, only holds for x in the interval $-\pi/2 \leq x \leq \pi/2$, even though $f^{-1}(f(x))$ is well defined for all values of x. Can you see why these two identities are so different? Enter $\sin^{-1}(\sin(x))$ as $Y1$, set $X_{min} = -20, X_{max} = 20, Y_{min} = -2 < -\pi/2, Y_{max} = 2 > \pi/2$, and press GRAPH to see the screen shown. What a dramatic difference between the graphs of these two identities!

Set $X_{min} = -2, X_{max} = 2$ and graph again to see the familiar line $Y = X$, but with corners at both ends corresponding to the points where X moves outside of the interval $[-\pi/2, \pi/2]$. Why does the graph behave this way? Zoom in on a corner to see that these points are indeed "sharp corners."

There are several ways to explain this graphical behavior. Look at the graph of $y = \sin(x)$. With a pencil, mark a point x_0 on the x-axis in the interval $[\pi/2, 3\pi/2]$ (an interval outside of the interval $[-\pi/2, \pi/2]$). Mark the correspond point $y_0 = \sin(x_0)$ on the y-axis, and follow that by marking the point $x_1 = \sin^{-1}(y_0)$ on the x-axis (which is now in the interval $[-\pi/2, \pi/2]$). Notice how x_1 decreases from $\pi/2$ to $-\pi/2$ as x_0 increases from $[\pi/2, 3\pi/2]$.

A more persuasive analytic argument involves a slope computation. If $f(x)$ is the function $f(x) = \sin^{-1}(\sin(x))$ then the *Chain Rule* implies that the slope of the graph is

$$f'(x) = \frac{1}{\sqrt{1 - \sin^2(x)}} \cos(x) = \frac{1}{\sqrt{\cos^2(x)}} \cos(x) = \frac{\cos(x)}{|\cos(x)|} = \pm 1$$

depending on whether $\cos(x)$ is positive or negative. This readily explains the shape of the graph.

Example 6.8 *The graph of the function $f(x) = \sin(\cos^{-1}(x))$ on the interval $[-1, 1]$ suggests a simplification for $f(x)$. State and prove an identity which simplifies $f(x)$.*

6.3. THE INVERSE TRIGONOMETRIC FUNCTIONS

Graphing $Y1 = \sin(\cos^{-1}(X))$ under ZoomFit with $X_{min} = -1, X_{max} = 1$ produces a graph which looks somewhat circular. In response to this, press [ZOOM] [5] to select the ZSquare plotting window. This confirms our suspicions. The graph appears to be the upper half of the circle of radius 1 centered at the origin. Is it the case that $f(x) = \sin(\cos^{-1}(x)) = \sqrt{1-x^2}$? To test our conjecture, we plot $Y2 = Y1 - \sqrt{(1-X\wedge 2)}$ under ZoomFit with $X_{min} = -1, X_{max} = 1$. This creates a "range" error,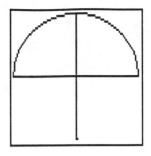
because Y_{min} and Y_{max} are both set equal to 0. This by itself is evidence in support of our conjecture. If Y_{max} is adjusted slightly to a small positive number, The graph of $Y \equiv 0$ appears, which supports our conjecture.

To prove our conjecture, write $y = \sin(\cos^{-1}(x))$ in the form $y = \sin(\theta)$, where $\theta = \cos^{-1}(x)$, and draw a right triangle showing this relationship between θ and x.

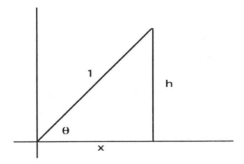

First write this equation in the equivalent form $\cos(\theta) = x$, and use this, and elementary properties of right triangle trigonometry to draw a right triangle with an adjacent side of x and a hypotenuse of 1 (see the figure on the next page). The *Pathagoreon Theorem* then implies that the opposite side h must have a value of $h = \sqrt{1-x^2}$. It follows that $y = \sin(\theta) = h = \sqrt{1-x^2}$, which proves the result. ∎

Example 6.9 *A 300 foot long ship whose bow is 57.2 miles east, and 37.6 miles north of an observer is steaming due west at 27 miles per hour, while the observer travels due north at 19 miles per hour. When the ship is within visual range, the observer plans to focus a telescope on the ship, adjusting the viewing angle of the lens so that it always fits exactly the 300 foot length of the ship. When is this viewing angle a maximum, and what is the distance between the observer and the bow of the ship at that time? When is the distance between the observer and the bow of the ship a minimum, and what is this distance?*

We begin by displaying in the figure below, all of the variables which play a role in the problem. Do we need all of these variables? Even if they serve only an intermediate purpose, an ample supply of variables can help immensely, and it is a good idea to display them promptly without being overly concerned about how they are to be used.

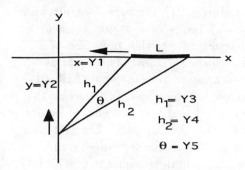

We wish to maximize the angle θ shown in the figure. Using the *Law of Cosines* we write

$$L^2 = h_1^2 + h_2^2 - 2h_1 h_2 \cos(\theta),$$

where L is the length of the ship, which must be expressed in miles to be compatible with other units. Solving the equation for θ gives us

$$\theta = \cos^{-1}\left(\frac{h_1^2 + h_2^2 - L^2}{2h_1 h_2}\right)$$

We wish to express $\theta = \theta(t)$ as a function of time t. Using the initial position and velocity of the ship and the observer we first write $x = x(t) = 57.2 - 27t$ and $y = y(t) = 19t - 37.6$, and then

$$h_1 = \sqrt{(x(t))^2 + (y(t))^2}, \quad h_2 = \sqrt{(x(t) + L)^2 + (y(t))^2}.$$

This gives us $\theta = \theta(t)$ as a rather complicated function of t, and the easiest way to enter it into our calculator is to enter all of the intermediate formulas as well. The calculator names $Y1, Y2, \ldots, Y5$ which will be used for this purpose are also shown in the figure. (*86* More natural letters can be used very effectively on this device. We will discuss this later. *86*)

Now let us enter this onto our calculator. From the home screen, enter 300/5280 and press [STO▶] [ALPHA] ([L]) [ENTER] to set L equal to the length of the ship in miles. We must use the variable X to denote the independent variable $t = time$. With this in mind, we enter:

$Y1 = 57.2 - 27 * X$ $\qquad$ $Y2 = 19 * X - 37.6$

$Y3 = \sqrt{(Y1 \wedge 2 + Y2 \wedge 2)}$ $\qquad$ $Y4 = \sqrt{((Y1 + L) \wedge 2 + Y2 \wedge 2)}$

$Y5 = \cos^{-1}((Y3 \wedge 2 + Y4 \wedge 2 - L \wedge 2)/(2 * Y3 * Y4))$

6.3. THE INVERSE TRIGONOMETRIC FUNCTIONS

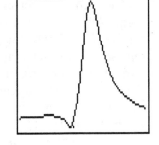

Deactivate all of the expressions except $Y5$, and set $X_{min} = 0$ (recall that X is time). To decide what we should use for X_{max}, consider the initial positions and velocities. Clearly the action will expire in less than 4 hours, and so we set $X_{max} = 4$. After an initial graph, we can see when all of the interesting activity is taking place. We set $X_{min} = 1.7$, $X_{max} = 2.4$ and selected ZoomFit to produce the screen shown.

To find the time $t = X$ for which $\theta = Y5$ is a maximum, we press [TRACE] and move the trace cursor to the approximate location $X = 2.09$ of the maximum.

For more accuracy, the slope equation $Y5'(X) = 0$ can be solved for $t = X$. An analytic formula for this derivative could be determined, but it would be very tedious to find it. Instead, we enter, in the SOLVER menu screen, the equation $eqn : 0 = fnDeriv(Y5, X, X, 10 \wedge -5)$, where the last entry is included to produce more accuracy. After moving the cursor to the line containing the unknown X, we press [ALPHA] ([SOLVE]) to get $t = X = 2.0939468$. We dropped several digits in the answer, and there is no simple way of determining how many digits are accurate. Using the $fnDeriv(\ )$ command has its disadvantages. (⋆86⋆ The $der1(\ )$ is more reliable. ⋆86⋆) It can be verified, however, that the answer is correct to as many decimal places as we have displayed. We press [X,T,θ,n] [STO▶] [ALPHA] ([A]) [ENTER] to store the answer as A.

An alternate way of getting the maximum value without using the SOLVER menu, is to use the $fMax$ command. From the home screen, press [MATH] [7] to select the command, and enter $fMax(Y5, X, 2, 2.2, 10 \wedge -5)$ to get $t = X = 2.09394$. We dropped the trailing digits, since they are beyond the accuracy of the method. The third and forth entries in this command specify the interval in which the the maximum value is to computed. The last entry in this command is, of course, the degree of accuracy of the computation.

The maximum angle is $\theta_0 = Y5(A) = .023634$ $radians = Y5(A) * 180/\pi = 1.35410°$. It occurs $t_0 = A = 2.09395$ $hours$ after the event began, and the distance between the observer and the bow of the ship at that time is $d_0 = Y3(A) = 2.2835$ $miles$. Notice that at time $t = t_0$ the observer will have passed the common intersection point where the origin of our coordinate system was placed, and will be heading away from this point while the ship approaches it.

The find the minimum distance between the observer and the bow of the ship we deactivate $Y5$, activate the distance function $Y3$, set $X_{min} = 0, X_{max} = 4$ and graph $Y3$ under ZoomFit. The graph has a clear minimum occurring near $X \approx 2.04$. The SOLVER menu is already set up from the first part of the problem. Just change $Y5$ to $Y3$ to get the equation $eqn : 0 = fnDeriv(Y3, X, X, 10 \wedge -5)$, which yields the solution $X = 2.07229$ after it is rounded off. We press [X,T,θ,n] [STO▶] [ALPHA] ([B]) [ENTER] to store this value as B. The minimum distance of $d_1 = Y3(B) = 2.1687 miles$ occurs $t_1 = 2.07229 hours$ after the initial moment in time. Notice that $t0 > t1$, so the distance is a minimum, before the viewing angle is a maximum. ∎

⋆⋆⋆⋆ 86 ⋆⋆⋆⋆

This problem evolves in a more natural way when more appropriate names are used. All of the names x, y, t, $h1$, $h2$, θ used above (**with one exception**) can be entered onto this calculator in the same way as they are defined above to represent expressions in time t. **The letter x is reserved by the calculator, and should not be used**, but there are few other restrictions. Names can be single letters, letters followed my numbers, or full words (Names must, however, start with a letter). From the home screen press [ALPHA] ([x]) [1] [ALPHA] ([=]) to enter $X1 =$. Follow this by entering the formula $X1 = 57.2 - 27 * t$, and then press [ENTER]. The calculator responds with "done". Press [2ND] [[alpha]] ([T]) to enter a small case letter t. Small case letters can be used throughout, but they require slightly more keyboard activity to enter. Enter the remaining expressions in the same way. The name θ can be entered by pressing [2ND] [[CHAR]] [F2] to access the Greek alphabet. After the expressions are entered, $\theta = \theta(t)$ becomes a function of t, and it cannot be graphed from the GRAPH menu. In order to graph $\theta(t)$ and solve the equation $\theta'(t) = 0$, we use instead the SOLVER menu. See the Ti-86 Guidebook.
★★★★ 86 ★★★★

6.4 Exponential Growth and Decay

The differential equation
$$\frac{dy(t)}{dt} = ky(t)$$
is so common in mathematics and its applications, that its solution
$$y(t) = Ce^{kt},$$
where C is a constant of integration, quickly becomes a familiar if not memorized formula to most students of mathematics. If the most common initial condition $y(0) = y_0$ is given, then C takes the form $C = y_0$. As much as any other problem, this classic differential equation justifies the introduction of the exponential function into the central core of mathematics.

Example 6.10 *A radioactive sample with a half life of 117.537 years must remain in protective storage until less than 0.005 grams of the substance are left. If 50 grams of the substance were put into a storage locker, how long will this take?*

Letting $y(t) = Ce^{kt}$ represent the amount of the substance present at time t, it readily follows that $C = 50$. If $t_h = 117.537$ denotes the half life, then there are $y(t_h) = 25$ grams present (half the original amount) when $t = t_h$. This gives us the equation $25 = 50e^{kt_h}$ which can be solved for k using a standard logarithmic argument.
$$25 = 50e^{kt_h} \Rightarrow \frac{1}{2} = e^{kt_h} \Rightarrow \ln(\frac{1}{2}) = \ln(e^{kt_h}) = kt_h.$$

This gives $k = -\ln(2)/t_h = -.0058972679$. **It is important not to round this number off to fewer decimal places.** Its position in the function gives it a pronounced affect on the values of the function. As soon as this number is entered onto the home screen, press [STO▶] [ALPHA] ([K]) [ENTER] to store the value as K.

6.4. EXPONENTIAL GROWTH AND DECAY

This function can now be graphed, although this is not entirely necessary. We enter $Y1 = 50 * e \wedge (K * X)$, Where X, of course, plays the role of time. It's easy enough to choose $X_{min} = 0$, but what should be use for X_{max}? The half life tells us that the 50 gram sample is reduced by half every 117 years, and so it seems reasonable to set $X_{max} = 2000$. We must find the value of X for which $Y1 = 0.005$. After the graph appears, press [TRACE] and move the trace cursor to the point having a y-coordinate of approximately 0.005. The corresponding x-coordinate (which is now the current value of X) is a good approximation. For more accuracy, press [MATH] [0] to select the SOLVER menu, and enter the equation $eqn : 0 = Y1 - .005$. Move the cursor the the line containing the unknown X (which already evaluates to our approximate solution), and press [ALPHA] ([SOLVE]) to get $X = 1561.797849957$.

This graphical approach is not the only approach, and arguably, it is not even the preferred approach. Using the model $y(t) = 50e^{kt}$, where $k = K$ is the value obtained previously, just set $y(t) = 0.005$ and solve, analytically, the resulting equation for t using the same logarithmic manipulations we used to find a value for k. From

$$.005 = 50e^{kt} \Rightarrow \frac{.005}{50} = e^{kt} \Rightarrow \ln\left(\frac{.005}{50}\right) = \ln(e^{kt}) = kt,$$

we get $t = \frac{\ln(0.005/50)}{k}$. Entered on the home screen, this value of $t = 1561.79785$ compares favorable to our previous value. ∎

Newton's Law of Cooling is a principle that leads to a differential equation very similar to the classic equation $\frac{dy}{dt} = ky$ just discussed, and it is frequently presented in calculus courses at this time. This principal states that the **rate at which a body cools is proportional to the difference between the temperature of the body and the temperature of the surrounding medium**. Translated into mathematics, the equation takes the form

$$\frac{dT(t)}{dt} = K(T(t) - A)$$

where $T(t)$ is the temperature of the body at time t, A is the constant temperature of the surrounding medium, and K is the constant of proportionality—a constant which depends on the physical properties of the medium.

This differential equation quickly turns into the well known equation $\frac{dy}{dt} = Ky$, if we just let $y(t) = T(t) - A$. As a consequence, we can easily specify the solution to *Newton's Law of Cooling* as

$$y(t) = T(t) - A = Ce^{Kt},$$

or in other words as

$$T(t) = A + Ce^{Kt},$$

where C is a constant of integration, whose value is determined by the initial condition.

Example 6.11 *A steel beam is taken out of a furnace and placed in an environment of $72°$ Fahrenheit. After 10 minutes the beam has a temperature of $1287°$, and after*

15 minutes, its temperature is 986°. If we must wait until its temperature is 100° before we can continue working with it, then how long must we wait? What was the temperature of the beam when it came out of the furnace?

Using the above equation as our model, we have $T(t) = 72 + Ce^{Kt}$. The temperature at $t = 10$ tells us that $T(10) = 1287 = 72 + Ce^{k10}$, or in other words that $1215 = Ce^{k10}$. In the same way, the temperature at $t = 15$ tells us that $914 = Ce^{k15}$. We cannot use our calculator to solve a system of two equations in two unknowns, but this is easy enough to do symbolically. We write

$$\frac{1215}{914} = \frac{Ce^{k10}}{Ce^{k15}} = \frac{e^{k10}}{e^{k15}} = e^{-5k},$$

so that

$$\ln\left(\frac{1215}{914}\right) = \ln(e^{-5k}) = -5k.$$

After entering $\ln(1215/914)/(-5)$ on the home screen, press $\boxed{\text{STO}\blacktriangleright}$ $\boxed{\text{ALPHA}}$ $\boxed{\text{K}}$ $\boxed{\text{ENTER}}$ to store the value of $k = -0.0569337569$ as K. Using either of the two equations, we can now evaluate C. Enter $1215/e \wedge (10 * K)$ and then press $\boxed{\text{STO}\blacktriangleright}$ $\boxed{\text{ALPHA}}$ $\boxed{\text{C}}$ $\boxed{\text{ENTER}}$ to store the value of $C = 2147.021742$ in memory. As we mentioned before, it is important to avoid rounding off the value for k since small differences in k can have significant impact. As long as it is stored in memory, there is no advantage to rounding it off.

Now that we have values for the two constants, the problem can be finished in the same way by solving the equation $100 = 72 + Ce^{kt}$ for t. Alternately, the equation $T = 72 + Ce^{kt}$ can now be graphed in the form $Y1 = 72 + C * e \wedge (K * X)$, and a value for X corresponding to $Y1 = 100$ can be found using the graph and the SOLVER menu. Using the first approach instead, we write $(100 - 72)/C = e^{kt}$ and take the logarithm of both sides to get $t = \ln((100 - 72)/C)/K = 76.22248475$. The beam will cool to 72° 76.222 *minutes* after it is taken out of the oven. It had a temperature of $T(0) = 72 + Ce^{k0} = 72 + C = 2219.021742°$ when it was first taken out of the oven . ∎

The study of differential equations is a vast and important branch of applied mathematics. We usually do not memorize solutions to differential equations as we did in this section, but we use instead, the theory and techniques presented in a course in differential equations taken soon after completing the calculus sequence.

★★★★ 86 ★★★★

A distinct advantage of this device over the TI-83 is that it can be used to solve some differential equations. The process starts by putting the calculator in its differential equation mode by pressing $\boxed{\text{2ND}}$ $\boxed{\text{MODE}}$ and selecting the Difeq mode. If you have access to this machine, and you are interested in the topic, you are invited to read the appropriate chapters in the TI-86 Guidebook.

★★★★ 86 ★★★★

6.5 The Hyperbolic Functions and Their Inverses

The hyperbolic functions and their inverses (some of them), are only available in the CATALOG menu. This menu is so big that it can be inconvenient to use, if you do not use the keyboard to scroll quickly to the desired command. Press [2ND] [CATALOG] and the **keyboard is automatically in alpha mode**. This makes it easy to choose a letter to go immediately to all the commands beginning with that letter. To enter the hyperbolic cosine onto the home screen, press (from the home screen) [2ND] [CATALOG] (c), use the ▼ key to scroll down to $\cosh^{-1}$ and press [ENTER]. (★86★ The hyperbolic functions and some of their inverses have a more convenient location on this device. Press [2ND] [MATH] [F4] to select the HYP submenu. ★86★)

Example 6.12 *Find a formula for* $\operatorname{sech}^{-1}(x)$ *which can be used on our calculator, and graph the function. Determine its domain analytically. Show analytically that* $\operatorname{sech}^{-1}(x) \to \infty$ *as* $x \to 0^+$. *Create a table of values of* $\operatorname{sech}^{-1}(x)$ *for* $x = 10^{-j}$ ($j = 1, 2, \ldots$).

This function is not available on our calculator, so the search for a formula has a certain practical value. From $y = \operatorname{sech}^{-1}(x)$, we write $x = \operatorname{sech}(y) = \frac{1}{\cosh(y)}$. This means that $\cosh(y) = \frac{1}{x}$, and consequently that $y = \cosh^{-1}(\frac{1}{x})$, or in other words

$$\operatorname{sech}^{-1}(x) = \cosh^{-1}\left(\frac{1}{x}\right).$$

The domain of $\cosh^{-1}(x)$ is (like any inverse function) the range of $\cosh(x)$, which is readily seen to be the interval $[1, \infty]$. Additionally, the domain of the hyperbolic cosine must be restricted to the interval $[0, \infty]$ in order to make this function one-to-one. It follows that that the domain of $\operatorname{sech}^{-1}(x)$ is the interval $[0, 1]$, and that its range is $[0, \infty]$. From the Y= menu, press [2ND] [CATALOG] (c) and scroll down to find the inverse hyperbolic cosine. Enter $Y1 = \cosh^{-1}(1/X)$, $X_{min} = 0$, $X_{max} = 1$, and graph it under ZoomFit to get the screen shown.

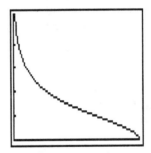

The above analysis also proves that

$$\lim_{x \to 0^+} \operatorname{sech}^{-1}(x) = \lim_{x \to \infty} \cosh^{-1}(x) = \infty,$$

but the graph does not show this very well. **It is worth reminding ourselves, that technology can be a valuable tool for the understanding and application of mathematics, but it must be rooted firmly in the more rigorous analytic aspects of the subject.**

We create a table of values to show how slowly $\operatorname{sech}^{-1}(x) \to \infty$ as $x \to 0^+$. In the Y= menu, set $Y2 = 10 \wedge (-X)$, and $Y3 = Y1(Y2)$. Deactivate $Y1$ so that it doesn't appear in our table. Press [2ND] [TBLSET], set TblStart= 1, ΔTbl= 1, press [2ND] [TABLE] and use the ▼ key to scroll through various parts of the table such as the screen shown. ∎

6.6 Indeterminate Forms and L'Hopital's Rule

We have now developed all of the so-called elementary special functions of mathematics, and they are frequently involved in complicated limit computations. This is an appropriate time to discuss a class of complicated limits called indeterminate forms. Limits of this sort, which carry descriptive labels such as

$$\frac{0}{0}, \ \frac{\infty}{\infty}, \ 0 \times \infty, \ \infty - \infty, \ 0^0, \ 1^\infty, \ \infty^0,$$

are defined in most standard calculus texts.

One might reasonably think that with a graphics calculator in hand, computing a limit should not be very difficult. Afterall, to evaluate a limit of a function $f(x)$, all one must do is zoom in on the appropriate point on the graph of $f(x)$ until the desired degree of accuracy is obtained. In practice, however, evaluating a limit in this way is not reliable. When the expression $\lim_{x \to a} f(x)$ is indeterminate, $f(x)$ becomes computationally unstable when x is close to $x = a$, and round off errors frequently become sizable. Recall what happen in Example 1.8 when we tried to evaluate a limit in this way.

If f, and g are differentiable functions on an open interval containing $x = a$, except possibly at a itself, and if $h = f/g$ is a $(0/0)$, or (∞/∞) indeterminate form at $x = a$, then, according to L'Hopital's Rule

$$\lim_{x \to a} h(x) = \lim_{x \to a} \frac{f(x)}{g(x)} = \lim_{x \to a} \frac{f'(x)}{g'(x)}$$

provided the limit on the right exists in a finite or infinite sense. Variations of this rule can be stated for left and right hand limits, and for limits with $a = \pm\infty$.

Example 6.13 *Compute the following limit both graphically and analytically.*

$$\lim_{x \to 0} \frac{6\sin(x) - 6x + x^3}{2x^5}$$

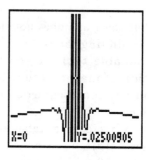

Enter $f(x) = (6\sin(x) - 6x + x^3)/(2x^5)$ as $Y1$ and graph it under ZoomFit with $X_{min} = -1$, $X_{max} = 1$. Place the graph cursor at the point on the graph corresponding to $x = 0$. (The trace cursor cannot be used, because the function is not defined at $x = 0$. After zooming in once, the graph begins to show signs of chaos near $x = 0$. Zoom in again, and the chaos near $x = 0$, shown in the screen, is quite dramatic. Here is a limit that we will not be able to evaluate graphically.

To evaluate the limit analytically, notice that the expression is clearly a $(0/0)$ indeterminate form, and so *L'Hopital's Rule* applies. After we apply the rule, we

6.6. INDETERMINATE FORMS AND L'HOPITAL'S RULE

have another (0/0) indeterminate form, and so *L'Hopital's Rule* can be applied again and again, until we arrive at a fraction, which is no longer a (0/0).

$$\lim_{x \to 0} f(x) = \lim_{x \to 0} \frac{6\cos(x) - 6 + 3x^2}{10x^4} = \lim_{x \to 0} \frac{-6\sin(x) + 6x}{40x^3}$$
$$= \lim_{x \to 0} \frac{-6\cos(x) + 6}{120x^2} = \lim_{x \to 0} \frac{6\sin(x)}{240x} = \lim_{x \to 0} \frac{6\cos(x)}{240}.$$

At this point, the limit is no longer a 0/0 indeterminate form and its value is obviously 1/40. ∎

Example 6.14 *Compute the following limit both graphically and analytically.*

$$\lim_{x \to 0^+} (1 + \sin(x))^{\cot(x)}$$

We proceed as we did in the last example. There is no button on the keyboard for $\cot(x)$, but this function can easily be entered in the form $1/\tan(x)$. The limit involved in this example is a 1^∞ indeterminate form, and typical of all indeterminates, the function itself is undefined at $x = 0$. The trace cursor gives no Y-value when $X = 0$.

After zooming in several times, the graph appears to remain stable at $x = 0$. This is encouraging! Move the cursor above and below the point on the curve "corresponding" to $X = 0$, and read the Y-values. Having already zoomed in several times, the first four digits of Y will have leveled off to 2.718 throughout the range of Y-values (a value close to e). Does this limit evaluate to e? It looks like it might, but if we zoom in again, our assessment might change, and if not, then perhaps it would if we zoomed in again. If we zoom in enough, then surely enough, round off error will become substantial, and the graph will lose its shape and become chaotic at the point.

The only verifiable way to evaluate this limit is to use *L'Hopital's Rule*. Before we can use the rule on a function f which has a 1^∞ indeterminate form, we must first write f in the form $f = e^{\ln(f)}$ and then consider the limit of $p = \ln(f)$. Using properties, of logarithms, p can then be expressed as a (0/0) or (∞/∞) form which is required before we can use *L'Hopital's Rule*. We write

$$f(x) = \exp\left(\ln((1 + \sin(x))^{\cot(x)})\right) = \exp\left(\cot(x) \ln(1 + \sin(x))\right),$$

and let

$$p(x) = \cot(x) \ln(1 + \sin(x)) = \frac{\ln(1 + \sin(x))}{\tan(x)}.$$

In this form, the function $p(x)$ is a (0/0) indeterminate as $x \to 0^+$, and so

$$\lim_{x \to 0^+} p(x) = \lim_{x \to 0^+} \frac{\frac{1}{1+\sin(x)} \cos(x)}{\sec^2(x)}.$$

At this point, the limit is no longer a 0/0 indeterminate form, and its value as $x \to 0^+$ is obviously 1. It follows that $f(x) = e^{p(x)} \to e^1 = e$ as $x \to 0^+$. ∎

6.7 Exercise Set

1. Show that $f(x) = 1+\log_2(x+2)$ and $g(x) = 2^{x-1}-2$ are inverse functions of each other. Demonstrate with an appropriate graph the symmetry between the curves. Choose a plot range of an appropriate size, and use the same scale on both axes.

2. Show that $f(x) = \frac{1}{300}x^3 - \frac{3}{200}x^2 + \frac{13}{400}x + 7$ is one-to-one for $\infty \leq x \leq \infty$. Create a table of values for the inverse function $g(x) = f^{-1}(x)$, for integer values of x between -50 and 50.

3. a) Show that $f(x) = x/\sqrt{x^2+4}$ is one-to-one on $(-\infty, \infty)$.
   b) Find a formula for the inverse function $g(x) = f^{-1}(x)$. What is its domain?
   c) Show graphically that $f(g(x)) = x$, for all x in the domain of g.
   d) Show analytically that $g((f(x)) = x$ for all real x.
   e) Demonstrate with an appropriate graph the symmetry between the curves $y = f(x)$ and $y = g(x)$.
   f) Show graphically that $g'(b) = \frac{1}{f'(a)}$ for $b = f(a)$ $(-\infty < a < \infty)$.

4. Show that $f(x) = x^3 - 6x^2 + 15x + 7$ is one-to-one on $(-\infty, \infty)$, so that $g(y) = f^{-1}(x)$ exists. Use the rule

$$g'(y) = \frac{1}{f'(x)} \quad (y = f(x))$$

to compute $g'(3)$ and $g'(y_0)$ for $y_0 = f(8)$

5. Find the largest interval on which $f(x) = x^2 - e^x$ is one-to-one. You may want to look at a graph for guidance, but don't rely on your eyesight. See how effectively you can use the differential calculus to establish this one-to-oneness. If $g(x)$ denotes its inverse, compute $g'(7)$.

6. Knowing that $\ln(x)$ is a logarithm allows us to determine its base without prior knowledge. Use the identity $\log_b(b) = 1$, and zoom in on the appropriate point on the graph of $y = \ln(x)$ to identify the base to 6 correct decimal places. Explain, without depending on the known value for e, why your answer has this degree of accuracy.

7. Evaluate the following. Notice their similarities or differences.
   a) $e^{3/2}$ b) $\sqrt{e^3}$ c) $\frac{1}{e^{\sqrt{2}}}$ d) $e^{-\sqrt{2}}$
   e) $\ln(7^2)$ f) $2\ln(7)$ g) $\ln(8+9)$ h) $\ln(8) + \ln(9)$
   i) $\ln(8 \times 9)$ j) $\ln(\frac{5}{3})$ k) $\ln(5) - \ln(3)$ l) $\frac{\ln(5)}{\ln(3)}$

10. Set up a procedure for estimating the value of $ln(x) = \int_1^x \frac{1}{t} dt$ with a Riemann sum obtained by dividing the interval $[1, x]$ into 40 subintervals of equal length and always using left end points of intervals to establish heights of rectangles. Use the procedure to estimate the value of $ln(0.1), ln(10), ln(100)$.

11. Use implicit differentiation to find the equation of the line tangent to the graph of $x^2 + y^2 = e^{x+y}$ at the point on the graph corresponding to $x = -1/2$. Is there more than one point on the curve corresponding to $x = -1/2$? Use a combination

6.7. EXERCISE SET

of graphical and and analytic ideas to show persuasively that there is only one such point on the curve.

12. Find the absolute maximum and absolute minimum, and where they occur, for the function f defined by
$$f(x) = \frac{5x^7}{e^x + e^{-x}}.$$
Include some evidence (not just visual evidence) that larger (or more negative) values will not be found outside of your plot window.

13. Evaluate the following exactly. Simplify your answers.
   a) $\log_7(16807)$ b) $\log_{16807}(7)$ c) $\tanh(\ln(e))$
   d) $\text{sech}(\ln(17))$ e) $\arctan(-\sqrt{3})$ f) $\sin(\arctan(17))$
   g) $\arcsin(\sin(\frac{78}{5}\pi))$ h) $\tanh(\text{arcsinh}(13))$

14. Without an antiderivative of a function $f(x)$, the *Fundamental Theorem of Calculus* can always be used to create an antiderivative. This is how (and why) we defined $\ln(x)$ earlier in this chapter, as an antiderivative of $1/x$. Suppose that the inverse trigonometric functions had not been introduced. Use the *Fundamental Theorem of Calculus* to define the antiderivative
$$T(x) = \int_1^x \frac{1}{1+t^2}\, dt$$
of $f(x) = \frac{1}{1+x^2}$. Numerical techniques (like *Simpson's Rule*, for example) are used to evaluate integrals, not antiderivative techniques.
   a) Create a table of values of $T(x)$ for $x = 1, 1.1, 1.2, \ldots$
   b) Compare the values of $T(x)$ to the values of $\arctan(x)$. Describe the relationship between them. (Both are antiderivatives of the same function. What does that imply?)

15. Evaluate each of the following:
   a) $\log_8(2.45)$ b) $\sqrt{\arctan(26)}$ c) $e^{\cos(5.3)}$ d) $\sqrt{2\sqrt{2}}$

16. Find the points of intersection of $y = x^3$ and $y = e^{x/600}$. Before you start looking, how do you know that there is more than one? A convincing analytical answer to this question is essential.

17. Evaluate the following integrals. Verify that the answers are correct, by evaluating the integral analytically, or by making an appropriate change in the variable of integration, and using the $fnInt$ command a second time.
   a) $\int_0^{5\pi} \frac{\cos(3x)\sin(3x)}{4+\cos^2(3x)}\, dx$ b) $\int_{-1}^{3} \frac{e^{2x}}{9+e^{4x}}\, dx$
   c) $\int_{-4}^{-2} \frac{x}{1-x^2}\, dx$ d) $\int_{-1}^{\sqrt{3}} \frac{1}{\sqrt{4-x^2}}\, dx$
   e) $\int_0^1 \frac{x}{\sqrt{4-x^2}}\, dx$ f) $\int_{-\sqrt{2}}^{1/e} \frac{x^3}{\sqrt{4-x^2}}\, dx$

18. Use a graphical means to verify the identity $\arctan\left(\frac{1+x}{1-x}\right) = \arctan(x) + \frac{\pi}{4}$. Verify the identity analytically (Hint: Use the idea that if if $F(x)$ and $G(x)$ are both antiderivatives of the same function, then they differ by a constant.)

19. Find the area (as a decimal number) of the region bounded by the curves

$$y = x^3, \text{ and } y = 5 - e^{-x}.$$

Verify the answer with an alternate calculation.

20. Find the area of the region bounded by the y-axis and the curves $y = \cos(x)$ and $y = -0.9 + \frac{3}{x-1}$. The two curves here actually bound infinitely many regions. Consider only the one region which is also bounded by the y-axis.

21. A movie theater has a 20 foot tall screen mounted on its front wall. The bottom of the screen is 20 feet above the floor, and the entire floor is flat and horizontal. How far from the front wall should a person sit in order to maximize the size of the viewing angle of the screen? What is the maximum viewing angle in degree measure?

22. When an environmental group first began to monitor the population of a threatened species of bird, there were an estimated 8,500,000 of the birds in the country. That was 50 years ago. The results of a current study suggest that the population has dwindled to 1,900,000. Emergency laws designed to save the bird from extinction are automatically placed in effect if and when the population drops to 500,000. Assume that over the lifetime of the study, the population has been subject to the same law of exponential decay. How much time remains before an emergency is declared?

23. The owner of a small pond is introducing bass into the pond for the first time. The pond is big enough and diverse enough to sustain a maximum population of 10,000 adult bass. It is expected that the population of adult bass will grow at a rate which is proportional to the difference between the maximum sustainable population and the current population. The pond is initially seeded with 50 adult bass. Three years later the population is estimated to be 1400. What will be the population be after fifteen years? How long will it take to reach a population which is 95% of it maximum population?

24. A certain bacteria in the human body is capable doubling its population every 45 minutes. If there were 20 bacteria in the initial contaminant, what will the population be 24 hours later? Assume that the population changes at a rate proportional to the size of the population.

25. In each of the following, try to compute the limit graphically, or by creating a table of values (if the limit exists). Create a window which demonstrates that the graphical approach is ultimately unreliable. Use *L'Hopital's Rule* to compute the limit analytically, or show that the limit does not exist.

a) $\lim_{x \to 1} \frac{2\cos(x-1) + x^2 - 2x - 1}{(x-1)^4}$ b) $\lim_{x \to \infty} (1 + \frac{2}{x})^x$

c) $\lim_{x \to 0^+} \sin(x) \ln(x)$ d) $\lim_{x \to 0^+} (\cos(5x) + \sin(2x))^{(5/x)}$

Project: Best seat in the theater

6.7. EXERCISE SET

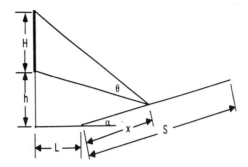

The above diagram describes the basic shape of a movie theater. A screen of height H feet is placed on a wall h feet above a level floor. The floor remains horizontal for the first L feet from that wall and then begins to rise at an incline of angle a as shown. This incline, of course, is where the seats are placed, and its length from the first row of seats to the last row in back is S. How far up this incline should we sit in order to have the best view of the movie? The best view, naturally, is one where θ is as large as possible.

Now suppose we plan to see a movie in a theater where $H = 30$, $h = 10$, $L = 20$, $S = 60$, $\alpha = 20°$, where the lengths are all in feet. Determine the value x which corresponds to a maximum viewing angle and find this angle θ. If the rows of seats are placed 3 feet apart with the first row right at the base of the incline, then what row should we sit in?

Suppose management plans to build a new theater where all of the dimensions would be kept the same except for α, which would be changed to $\alpha = 16.5°$. Determine how much this would change the viewing angle by computing the new values of x and θ.

Chapter 7

Sequences and Series

Imagine the large supply of functions we can create and use with our current knowledge of mathematics. This collection would include the power function, $p(x) = x^r$, where r is a real constant, all of the polynomial and rational functions, the trigonometric, logarithmic, exponential, hyperbolic functions and all of their inverses. It would include all of the combinations of these functions using finitely many compositions and finitely many algebraic operations. These are the functions, which we have the power to know. We may need our calculators's help, but we can evaluate them, graph them, manipulate them, and study their behavior.

The collection is large, and we can certainly use these functions to create a variety of complex mathematical models. Yet, one of the truly fascinating aspects of mathematics is that this supply of accessible functions occupies only a small corner in a much larger warehouse of "nice," "well-behaved" functions that play a significant role in mathematics and its applications. Most of the functions in this warehouse live in a world that is currently beyond our reach.

Examples come from a variety of mathematical situations. We have already mentioned that antiderivatives of continuous functions are frequently (actually, usually) beyond our reach, even though the *Fundamental Theorem of Calculus* says that they always exist. The same can be said about solutions (of the form $y = f(x)$) to some fairly simple equations in x, and y. Inverses to some functions, even simple functions like $f(x) = x + \sin(x)$, for example, can often be shown to exist (this is usually easy to show), but formulas for them are frequently (no, usually) beyond our reach. Solutions to differential equations, which are so important to mathematical applications, are frequently hard to describe.

In all of these situations, and others as well, answers to very practical problems can turn out to exist, yet be beyond our ability to describe in terms of familiar functions. How are we to reach these functions, and study their properties? Certainly this is an important issue, if it can be addressed.

In this chapter we study ideas, which will ultimately provide us with a means to get a handle on all of the functions we usually need in mathematics, including those that may be quite inaccessible otherwise. It turns out that functions in this list can be described, surprisingly enough, by expressions that look and behave very much like the simplest functions, namely polynomials

$$p(x) = \sum_{j=0}^{n} a_j x^j, \text{ or } p(x) = \sum_{j=0}^{n} a_j (x-a)^j.$$

The only difference is that we must allow the degree of the polynomial to go to infinity. In the process, "long polynomials"

$$p(x) = \sum_{j=0}^{\infty} a_j x^j, \text{ or } p(x) = \sum_{j=0}^{\infty} a_j (x-a)^j$$

called **power series**, and **Taylor series** are created.

Realizing a function as a power series gives us an important means to study the behavior of the function. It turns out that power series behave very much like ordinary polynomials, although this requires a careful analysis. Frequently, the coefficients, a_n ($n = 1, 2, \ldots$), for a particular power series can all be evaluated, and this gives us a very specific way to formulate the function represented by the power series.

This is not an easy topic, and there are many issues in this chapter that must be discussed along with power series. Power series and Taylor series are introduced in the last section in this chapter, after all the other details are considered. Since it is the most important issue, some text books prefer to discuss power series first, especially Taylor series, followed by all of the other details. In that case, read a portion of the last section of this chapter first. It may help to postpone the material on power series until later, since this is a bit more abstract.

7.1 Sequences

While the concept of a sequence is needed for the development of the rest of the chapter, it is also a concept, which is interesting in its own right. Intuitively, a sequence is just an ordered list of real numbers, which in this chapter will always be an infinite list. A formal definition, however, is more useful in an environment where technology is used. A sequence, formally, is just a function whose domain is the set of of positive integers, or a subset of that set. Suppose, for example, that $\{a_n\}$ denotes the sequence (function),

$$\left\{ \frac{(-1)^n}{n^2} \right\}.$$

There are several ways to enter this sequence on our calculator. A sequence is, after all, just a function, and so the standard way of entering a function can be used. We can also use the $seq(\ )$ command, which was used extensively in an earlier chapter. Enter $Y1 = (-1) \wedge X / X \wedge 2$ in the Y= menu. We compute $a_{25} = Y1(25)$ in the usual way by going to the home screen and pressing [VARS] ▶ [1] [1] to select Y1. Finish with $Y1(25)$ and press [ENTER], to get $a_{25} = Y1(25) = -.0016$. Recall that a decimal can be expressed in fractional form as well. Press [2ND] [[ANS]] [MATH] [1] (to select ▶Frac) [ENTER] to get $a_{25} = -1/625$.

7.1. SEQUENCES

To create a table, press $\boxed{\text{2ND}}$ $[\![\text{TBLSET}]\!]$, set TblStart= 1 and ΔTbl= 1. Press $\boxed{\text{2ND}}$ $[\![\text{TABLE}]\!]$ to see the table of values of $a_n = Y1(X)$ $(n = X)$. As we scroll down the table, it becomes clear that $\lim_{n\to\infty} a_n = 0$. **Setting ΔTbl equal to a large number is one way of scrolling through table values at a faster rate of speed.** It is, however, only one way of letting $n \to \infty$ quickly, and we will discuss ways in the examples of speeding up the process even more.

A sequence (since it is just a function) can usually be graphed, but **this particular sequence (function) cannot be graphed**. This is more a consequence of the formula used to define the sequence rather than any more fundamental barrier. As long as $X = n$ is an integer, the value of $a_n = Y1(X)$ is well defined. However, $(-1)^X$ (and hence $Y1(X)$) is undefined for most values X in any plotting range, which naturally includes all of the real numbers in the range.

A sequence can also be entered formally as a sequence (a list) on our calculator. We used the $seq(\ )$ command extensively to compute Riemann sums in an earlier chapter (**see page 47 for a review**). A sequence on our calculator, produced with the $seq(\ )$ command, is a list of numbers enclosed by set brackets $\{\ldots\}$. A list (sequence) can contain at most 999 elements. Alas, we can only enter finite sequences in this way. To enter the first 500 elements of the above sequence from the home screen, press $\boxed{\text{2ND}}$ $[\![\text{LIST}]\!]$ ▶ $\boxed{5}$ to select the $seq(\ )$ command. Enter $seq((-1)\wedge N/N\wedge 2, N, 1, 500)$ and press $\boxed{\text{STO▶}}$ $\boxed{\text{ALPHA}}$ $([\![A]\!])$ $\boxed{\text{ENTER}}$ to store the sequence as A. Recall that to access our newly created sequence A, the prefix L must be attached to the letter A. This prefix appears in the OPS submenu of the LIST menu, but there is an easier way to attach this prefix to the name of a sequence. The letter A automatically appears in the NAMES submenu of the LIST menu. From the home screen, press $\boxed{\text{2ND}}$ $[\![\text{LIST}]\!]$, select A through the NAMES submenu, press $\boxed{\text{ENTER}}$ and the list LA will appear on the home screen with the prefix attached. To compute a_{25}, enter L$A(25)$ and press $\boxed{\text{ENTER}}$.

Another interesting (and very different) way of entering a sequence is to change a mode setting which fundamentally alters the way our calculator accepts functions. Press $\boxed{\text{MODE}}$ and scroll over to the far right hand side of the fourth line to select **Seq**, and press $\boxed{\text{ENTER}}$. This changes the Y= menu to a form which is particularly suitable for sequences. (★86★ The Seq mode is not available on this device. ★86★) The $\boxed{\text{X,T,}\theta\text{,n}}$ key now represents the variable n rather than X. Press $\boxed{\text{Y=}}$ and notice the change in the menu into a form for defining sequences. Let us disregard the terms in this menu having no impact on our current example, and scroll to the line which displays $u(n) =$. Enter $u(n) = (-1)\wedge n/n\wedge 2$. For example, to evaluate a_{25}, go to the home screen, and press $\boxed{\text{2ND}}$ $[\![u]\!]$ to enter the sequence u onto the input line. **Notice the buttons which were pressed.** The u we selected is a gold u, not a green u. It appears above the $\boxed{7}$ key, and to select it, we **pressed the** $\boxed{\text{2nd}}$ **key, not the** $\boxed{\text{ALPHA}}$ key. Follow this with $u(25)$ and press $\boxed{\text{ENTER}}$ to get $a_{25} = u(25) = -.0016$.

A graph or table can be created for the sequence $u(n) = \{a_n\}$. Either one can be used to suggest a probable value for $\lim_{n\to\infty} a_n$ (which is clearly 0 in this simple example).

To create a graph Press $\boxed{\text{WINDOW}}$ and notice all of the parameters which need to be set. The values for $X_{min}, X_{max}, Y_{min}, Y_{max}$ still control window size, and the new parameters nMax and nMin have an expected meaning. Press [GRAPH] to see a plot of all the ordered pairs $(n, u(n))$ selected from the WINDOW menu.

To create a table, press [2ND] [TABLE], set TblStart= 1 and ΔTbl= 1. Press [2ND] [TABLE] to see the table of values of $a_n = u(n)$. As we scroll down the table, we see again that $\lim_{n\to\infty} a_n = 0$.

The "Seq" mode should be ideally suited for our work with sequences (one would think), but it is not. It takes much more time to perform a computation with $u(n)$ than the same computation takes with $Y1(X)$. As long as a_n is defined <u>explicitly</u> as a formula in n, we will use $Y1(X)$ to define a sequence. A sequence can also be defined <u>recursively</u> (not as a formula in n) and in this case the notation $u(n)$ available in the "Seq" mode is an excellent way to define a sequence. We will introduce recursively defined sequences shortly, and that will be an appropriate time to discuss why computations involving $u(n)$ are so slow.

Example 7.1 *Show that the sequence $\{a_n\}$ defined by $a_n = \frac{n^{100}}{100^n}$ is eventually a decreasing sequence which converges to zero. At what point in the sequence does it become decreasing? Determine an N such that $0 < a_n < 10^{-8}$ for all $n \geq N$.*

Press [MODE], return the calculator to the "Func" mode, and enter $Y1 = X \wedge 100/100 \wedge X$ in the Y= menu. We begin by evaluating, for the sake of interest, the first few terms of the sequence. We can evaluate the first 9 terms of this sequence, but any attempt to evaluate $Y1(X)$ for $X \geq 10$ will generate an OVERFLOW ERROR. We can see why by looking at the values of the first 9 terms.

$$0.01,\ 1.3\,10^{26},\ 5.2\,10^{41},\ 1.6\,10^{52},\ 7.9\,10^{59},\ 6.5\,10^{65},\ \ldots,\ 1.0\,10^{81}$$

Does this sequence converge to infinity? Surprisingly enough, it does not, but the OVERFLOW ERROR continues even though the sequence converges to zero. While the value of the fraction is ultimately well within the numerical bounds of our calculator, the numerator and denominator are simply too large.

One way to bring down the size of these numbers is to work instead with $\ln(a_n)$, which is much smaller than a_n. Using the common logarithm might be even more convenient, since it would be easier to view the value of a number by looking at its base 10 logarithm. For example,

$$\log_{10}(a) = 14.73 \Rightarrow a = 10^{\log_{10}(a)} = 10^{14.73} = 10^{.73}\,10^{14} \approx 5.37\,10^{14}.$$

This allows us to see at a glance what the value of a is, by looking at the value of $\log_{10}(a)$.

Before we enter the sequence $\{\log_{10}(a_n)\}$ on our calculator, it must be simplified, or we will continue to generate OVERFLOW problems with large numbers. We write.

$$\log_{10}(a_n) = \log_{10}\left(\frac{n^{100}}{100^n}\right) = \log_{10}(n^{100}) - \log_{10}(100^n) = 100\log_{10}(n) - n\log_{10}(100).$$

Using this simplification, enter $Y1(X) = 100\log(X) - X\log(100)$. Press [2ND] [TBLSET] and set TblStart= 1 and ΔTbl= 10 (a number larger than 1 just to scroll

7.1. SEQUENCES

through sequence values faster). Press [2ND] [[TABLE]] and observe sequence values. Remember that they represent logarithms of our primary sequence values. Using the above analysis, the value of $Y1(21) = 90.222$, for example, means that a_{21} is the very large number $a_{21} = 10^{0.222}\, 10^{90}$. The values of a_n are decreasing by the time that $n = 31$. Scroll further down the table to get the screen shown, where the values appear to be tending to $-\infty$. The value, for example, of $Y1(141) = -67.08$ means that a_{141} is the very small number $a_{141} = 10^{-0.08}\, 10^{-67} = 0.83 \cdot 10^{-67}$. This is solid evidence that the sequence $\{a_n\}$ converges to 0.

X	Y1
81	28.849
91	13.904
101	-1.568
111	-17.47
121	-33.72
131	-50.27
141	-67.08

X=141

At what point does the sequence turn around and begin to decrease? Our table clearly indicates that it turns around in the interval $11 < X < 31$. This is close enough that we could reset ΔTbl to ΔTbl= 1, and get the value exactly. Using this approach, the sequence decreases after its largest value of $Y1(22) = 90.242$. **If this high point were more obscure, we might be able to find it by solving the equation $Y1'(X) = 0$ for X.**

We can see from the above screen that a_n will decrease below 10^{-8} somewhere in the interval $101 < n < 111$. In the TBLSET menu, keep ΔTbl= 1, set TblStart= 101, and press [2ND] [[TABLE]]. The values of $Y1(105) = -7.881$ and $Y1(106) = -9.469$, mean that $a_n < 10^{-8}$ for $n \geq N = 106$. ■

Can we trust the above calculation? What can we say analytically, without too much effort? Notice that $\lim_{n \to \infty} a_n$ is a ∞/∞ form. If we apply L'Hopital's Rule to a_n, the numerator will become $100n^{99}$ and the denominator will be somewhat unchanged at $\ln(100)100^n$. This process continues, of course, with repeated applications of L'Hopital's Rule, and after 100 applications of the rule, we will have a limit which clearly goes to 0.

In closing, let us remember that merely listing the first 9 terms, and witnessing the phenomenal growth of this sequence, did not suggest the long term pattern of values. The sequence converges to 0. There is a sequence $\{a_n\}$ in the exercises, where a_n decreases to essentially 0 for the first N terms, where N is an incomprehensibly large number, yet the sequence ultimately begins to increase and diverges to ∞.

Example 7.2 *Show that the sequence $\{a_n\}$ defined by $a_n = \left(\frac{n+3}{7+n}\right)^{2n}$ converges, and find its limit. Verify its limit by an alternate means.*

Our first impulse might be to reach for our calculator, but before we do, we should think about what we are asking our calculator to do. For example, $a_{200} = \left(\frac{203}{207}\right)^{400} = (.9806763285)^{400}$. Can we trust technology on this calculation? As we multiply $b = 0.9806763285 \approx 1$ by itself over and over again, the product will clearly stay close to 1 for some time, but will the product get close to 0 or 1 or some number in between, by the time we multiply b by itself 400 times? **This limit is an example of a 1^∞ indeterminate form.** It is difficult to predict its value by observation, and hard to decide how well it will be handled by our calculator.

We should be skeptical as we enter $Y1 = ((X + 3)/(7 + X)) \wedge (2 * X)$, set TblStart= 1, ΔTbl= 10 in the TBLSET menu, and press [2ND] [[TABLE]]. Our calculator seems to have no trouble generating values. As we scroll down the table, the values continue to decrease, but they do not seem to level off. Does this sequence converge to 0? We need to scroll further down the table to a point where the values level off before we can make any conclusions about the value of the limit. We could get $X \to \infty$ faster by setting $\Delta\text{Tbl}= 100$ or even 1000, but that doesn't seem to help. We can, however, force X to get large at a much faster rate. To this end, set $Y2 = 2 \wedge X$ (for example), and set $Y3 = Y1(Y2)$. Deactivate $Y1$ so we don't see its values. Set TblStart= 1 and $\Delta\text{Tbl}= 1$. Essentially, this set up creates a table of values $Y1(X)$ where X doubles in size in each table entry. Press [2ND] [TABLE] and scroll through the table entries to get the screen shown. As you can see, it takes a while for $Y2 = 2^X$ to get large, but eventually it gets large at a rate which is much faster than X itself regardless of how large a value given to ΔTbl. The entries seem to level off, and if our calculator is to be trusted, it appears that $\lim_{n \to \infty} a_n = 3.4 \, 10^{-4} = 0.00034$, a small number, but definitely nonzero. Greater accuracy will be displayed if a value is computed from the home screen. For example, $Y3(40) = 0.0003339749225$.

Enter $Y3(50)$, on the other hand, and we get $Y3(50) = 1$. Every indeterminate form, $\lim_{x \to x_0} f(x)$, is **ultimately unstable.** All we have to do is choose x close enough to x_0. In the present case, it is easy to see why. Enter $2 \wedge 50$ and press [STO▶] [X,T,θ,n] [ENTER], to store this value to X. Enter $(X + 3)/(7 + X)$, press [ENTER], and the value we get (rounded off to 10 decimal places) is exactly 1. Raise this to the power 2^X and, of course, a value of 1 is returned again.

To gain a deeper understanding of this limit, we could use $L'Hopital's$ Rule to verify its value. The standard approach calls for us to first rewrite the expression $a(n)$ as $\exp(\ln(a(n)))$, and then to consider the limit of $\ln(a(n))$. This step is omitted, but further insight into this kind of calculation can be found by reviewing our work with $L'Hopital's$ Rule in Example 6.13, and Example 6.14. Using this technique on a sequence $\{a_n\}$ is no different than using it on a function $f(x)$. One simply treats the variable n as a continuous variable similar to the variable x.

Actually, the "standard approach" from the last paragraph suggests an alternate way of entering an equivalent sequence on our calculator. Write

$$b_n = \ln\left[\left(\frac{n+3}{7+n}\right)^{2n}\right] = 2n \ln\left(\frac{n+3}{7+n}\right) = 2n\ln(n+3) - 2n\ln(7+n).$$

This sequence, an $\infty - \infty$ indeterminate form, inspires more confidence, but it is still an indeterminate form. Enter $Y1 = 2 * X * \ln(X + 3) - 2 * X * \ln(7 + X)$, and deactivate $Y1$, so that it doesn't appear in our table. (We have already set up $Y2$ and $Y3 = Y1(Y2)$, and they can be used again.) Set TblStart= 1, $\Delta\text{Tbl}= 1$, press [2ND] [TABLE] and scroll through the values to get the screen shown, where the values have leveled off to -8. This suggests that $\lim_{n \to \infty} b_n = -8$, which implies that

7.1. SEQUENCES

$\lim_{n\to\infty} a_n = e^{-8}$. To see if this is the same value as our first answer, we compute $e^{-8} = 0.000354626279$. ∎

We should add, however, that $Y3(40)=0$. Alas, the computation problems of an indeterminate form are inescapable. Computing the limit analytically, using *L'Hopital's Rule* is the only reliable way to get a value. It turns out that $\lim_{n\to\infty} a_n = e^{-8}$ is the correct answer.

The sequences that we have worked with so far have all been explicitly defined. In other words, they have all been of the form $\{a_n\}$ where a_n is given as a formula in n. A sequence can also be defined by expressing the nth entry in the sequence in terms of some (or all) of the previous $n-1$ entries of the same sequence. Such a sequence is said to be **recursively defined**. It is a more natural way of defining some sequences.

Example 7.3 *Determine the sequence $\{b_n\}$ of monthly balances corresponding to a loan of \$20,000, borrowed at a 9% annual interest rate compounded monthly, with a monthly payment of \$400. What is the balance after four years.*

One of the most interesting features of this example is the easy way it can be formulated. **There is no need to look up a financial formula.** Based on the simple idea that

$$\text{new balance} = \text{previous balance} + \text{interest} - \text{payment}$$

we simply write

$$b_{n+1} = b_n + \frac{.09}{12} b_n - p = (1 + \frac{.09}{12}) b_n - 400, \quad b_0 = 20000.$$

This defines the sequence $\{b_n\}$ recursively. Knowing $b_0 = 20000$, we can use this formula (with $n = 0$) to compute b_1, and then, in turn, $b_2, b_3, \ldots$.

The "Seq" mode is ideally suited for computations involving recursively defined sequences. (★86★ The Seq mode is not available on this device.★86★) Press MODE and scroll over to the far right hand side of the fourth line to select **Seq**, and press ENTER. The Y= menu is now in a form which is particularly suitable for entering a recursively defined sequence. Recall that the X,T,θ,n key now represents the variable n rather than X. Enter $u(n) = (1 + .09/12) * u(n-1) - 400$ in the Y= menu. Recall that the symbol u appears above the 7 key, and it is entered by pressing 2ND [u] (not ALPHA [U]). SET $nMin = 1$ and $u(nMin) = 20000$. Create a table in the usual way with TblStart= 1 and ΔTbl= 1. Press 2ND [TABLE] to see the values of $u(n)$, or to simply find the balance after four years, go to the home screen and press 2ND [u] (4 8) ENTER to get $u(48) = \$5619.82$. Values are also displayed more accurately on the home screen.

Notice how slow the computations are. The larger the value of n, the more time it takes to compute $u(n)$. This is because each call to evaluate $u(n)$

evaluates $u(j)$ as well for all $j \leq n$. This sequential mode is very convenient, but if n is large or if speed is important, **a better way to enter a recursively defined sequence is to write a program.** (★86★ Without a Seq mode, writing a program is the only way to enter a recursively defined sequence. ★86★) It may help to consult Sec 6.1, where a program for an inverse function was created. It can be used as an example of how to enter, edit, and run a program. The <u>Ti 8x Guidebook</u> is also an excellent reference.

The following program computes a sequence of balances for a loan of A dollars, borrowed at I% annual interest compounded monthly (entered as a percent), with a monthly payment of P dollars.

PROGRAM: LOAN	This names the program
:Disp "AMOUNT"	
:Input A	
:Disp "INTEREST"	
:Input I	
:Disp "PAYMENT"	
:Input P	
:Delvar($\text{L}B$)	This deletes any previous value of the sequence B.
:$A \to \text{L}B(1)$	This sets the first sequence value.
:$1 \to J$	
:Lbl Q	This starts the program loop named Q
:$(1 + I/1200) * \text{L}B(J) - P \to \text{L}B(J+1)$	
:While $\text{L}B(J+1) > 0$	A condition.
: $J + 1 \to J$	A command under *condition*.
:Goto Q	Another command under *condition*.
: End	This marks the end of the "While" group.
:Disp "output=B"	A reminder of where to find the output
	Look for (=) in the TEST menu
	(★86★ Look for (=) on the keyboard. ★86★)
: Stop	

When this program is run, The only output is a reminder that the answer is a sequence called B. One interesting way (among many) to display the sequence B of balances is to **display it on the STAT EDIT screen.** Press [STAT] [5] to select SetUpEditor. Enter SetUpEditor $\text{L}B$ and press [ENTER]. This clears the STAT EDIT screen and places the sequence B of monthly balances in the first column of the STAT EDIT screen. To see the display of values, press [STAT] [1] to select the EDIT screen.

It should be apparent, when this program is run and the results are viewed, that a substantial increase in computational speed has been realized. It is easy to see why. Each term $B(J)$ is computed only once. ∎

7.2. SERIES

★★★★ 86 ★★★★

Programs are easier to write on this device, and as we have mentioned before, there is no "L" prefix to attach to the name of a list. After the program is run, we can display the list on the LIST EDIT screen or the STAT EDIT screen (they are the same). Press [2ND] [LIST] [F5] to select the OPS submenu, press [MORE] [MORE] [MORE] [F3] to select SetLE (which plays the same role as SetUpEditor on the Ti83), enter SetLE B and press [ENTER]. This clears the LIST EDIT screen and places the sequence B in the first column. To see the sequence of monthly balances, press [2ND] [M4] (from inside the OPS submenu) to select the EDIT screen.

★★★★ 86 ★★★★

Example 7.4 *A very well-known sequence, called the Fibonacci sequence, is defined recursively by*

$$x_{n+2} = x_{n+1} + x_n, \ x_1 = x_2 = 1.$$

Create a table of values of the Fibonacci sequence. Determine the 40th term, x_{40}.

This sequence was first formulated 800 years ago to describe the populations of successive generations of breeding rabbits. Since that time, it has found its way into many mathematical problems. In addition, the Fibonacci sequence is found as a number pattern in a surprisingly large number of diverse settings in nature. It is such an interesting sequence that a mathematical journal, <u>The Fibonacci Quarterly</u>, is devoted to its study.

In this example, x_n depends on its two previous values. Consequently, x_1 and x_2 must both be given as start up values, before we can compute x_n for $n \geq 3$. **Notice how these start up values are entered** in the Y= menu.

Press [MODE], select the Seq mode and press [ENTER]. Press [Y=] and enter $u(n) = u(n-1) + u(n-2)$, $nMin = 1$, $u(nMin) = \{1, 1\}$. Curly brackets are used to enclose the values $x_1 = 1$ and $x_2 = 1$. Set TblStart=1, ΔTBl= 1, and press [2ND] [TABLE] to see the sequence of values, which is not included. Values are displayed with more accuracy from the home screen. Press [2ND] [u] [(] [4] [0] [)] [ENTER] to get $x_{40} = u(40) = 102334155$. ∎

7.2 Series

To say that a series is an infinite sum **misses an important part of the definition** which makes the concept of a series much more comprehensible. Fortunately, graphing calculators can be used very effectively to study this important definition.

A series $\sum_{j=0}^{\infty} a_j$ is a sequence—**a sequence**—a sequence $\{s_n\}$ of partial sums. Each term s_n in this sequence is much more manageable than the whole series, because each term $s_n = \sum_{j=0}^{n} a_j$ is a <u>finite</u> sum. We already have a good understanding of finite sums, and the previous section gave us a background in sequences. So now our attention will be focused on this sequence $\{s_n\}$. Does it converge or diverge? If it converges, can we compute its limit? In the next example, we use our calculator to shift our attention from the series (as an infinite sum) to its sequence of partial sums.

Example 7.5 *Compute the sequence of partial sums for the series*

$$\sum_{j=1}^{\infty} \frac{1}{j^2}.$$

Assuming that the series converges, approximate its sum by computing the sum of the first 500 terms, and the sum of the first 2000 terms.

There are several ways to create this sequence, with advantages and disadvantages to each approach.

In the last section, we used the Y= menu to create a sequence, and it can be used in the same way to create a sequence of partial sums. Press [Y=], and after clearing any previously defined function for $Y1$, press [2ND] [LIST] ▶ ▶ (to select the MATH submenu) [5] (to select sum( )). Follow this by pressing [2ND] [LIST] ▶ (to select the OPS submenu) [5] (to select seq( )). Finish by entering

$$Y1 = sum(seq(1/J \wedge 2, J, 1, X)).$$

This creates the sequence $\{s_n\}$ of partial sums

$$s_n = \sum_{j=1}^{n} \frac{1}{j^2} = Y1(X) \, for \, X = n.$$

We create a table of values in the usual way. Press [2ND] [TBLSET], set Tbl= 1 and ΔTbl= 1. If we wish to scroll more quickly through the values of the sequence $\{Y1(X)\}$, we can give ΔTbl a larger value (we let ΔTbl= 10 to produce the screen shown), or we can use some of the methods of the previous section to speed up the scrolling process even more. Press [2ND] [TABLE] and scroll down until the values begin to stabilize somewhat at approximately 1.64 to get the screen shown. From the home screen enter $Y1(500)$ to get (after a 15 second wait) $s_{500} = Y1(500) = 1.642936066$.

Shall we try $Y1(2000)$ in the same way? It doesn't work. The sum command operates on a "list" (the list of individual terms in the sum produced by the seq( ) command) and a list can have a maximum of 999 entries. It follows that this method can only be used to compute the first 999 terms of the sequence of partial sums.

It is also evident as we scroll down through the terms $Y1(X)$ that it takes considerably more time to compute $Y1(X)$ as X increases. Each evaluation of $Y1(X)$ computes the entire sum from $j=1$ to $j=X$, and naturally this takes more time when X is large. This is surely an inefficient way to compute the sequence of partial sums when speed is a critical issue. For example, once we have a value for $Y1(249)$, we don't have to add all 250 terms in the sum to compute $Y1(250)$. All we have to do is take advantage of the simple observation that

$$Y1(250) = Y1(249) + \frac{1}{250^2}.$$

7.2. SERIES

To compute a sequence of partial sums in this more efficient way, we write a program. The following program generates the sequence $\{s_k\}_{k=m}^{k=n}$ of partial sums for the series $\sum_{j=m}^{\infty} a_j$, where

$$s_m = a_m, \; s_{m+1} = a_m + a_{m+1}, \; \ldots, s_n = \sum_{j=m}^{n} a_j.$$

It assumes that the terms $\{a_j\}$ in the sum are entered in the Y= menu in the form $Y1 = a_j$ with $j = X$.

PROGRAM: SERIES
:Disp "FIRST"   Enter the value of m at this prompt.
:Input M
:Disp "LAST"   Enter the value of n at this prompt.
:Input N
:DelVar LS
:$M \to X$
:$Y1 \to LS(1)$
:Lbl Q
:L$S(X) + Y1(X+1) \to LS(X+1)$
:IS> (X, N)
:Goto Q
:Disp "OUTPUT=S"   A reminder of where the output is.
                                  Look for (=) in the TEST menu
                                  ($\star$86$\star$ Look for (=) on the keyboard. $\star$86$\star$)

Before we run this program, we must set $Y1 = 1/X \wedge 2$ in the Y= menu. To see the sequence of partial sums after the program is executed (we are reminded that the output is a list S), press $\boxed{\text{STAT}}$ $\boxed{5}$, enter SetUpEditor LS, and press $\boxed{\text{ENTER}}$ to put the list S on the STAT EDIT screen. Press $\boxed{\text{STAT}}$ $\boxed{1}$ to view the sequence. With this program, the sequence of partial sums will compute quickly and efficiently. However, since the output is in the form of a list S, we are still limited to placing at most 999 entries in S.

If we wish to create sums of more than 999 elements, we must avoid altogether the construction of large lists. The following program simply computes the numerical answer for the final sum, namely $\sum_{j=m}^{n} a_j$. As in the last program, We must enter $Y1 = a_j$ with $j = X$ in the Y= menu.

PROGRAM: BIGSUM
:Disp "FIRST"   Enter the value of m at this prompt.
:Input M
:Disp "LAST"   Enter the value of n at this prompt.
:Input N
:$M \to X$
:$Y1 \to S$
:Lbl Q

```
:S + Y1(X + 1) → S
:IS> (X, N)
:Goto Q
:Disp "S="                    This prints S=.
:Disp S                       This prints the answer.
```

Run this program with FIRST=1 ($m = 1$), and LAST=2000 ($n = 2000$), to obtain, after approximately $2\frac{1}{2}$ minutes of run time an answer of

$$S = \sum_{j=1}^{2000} \frac{1}{j^2} = 1.644434442.$$

Surely, this value must be close to the corresponding infinite sum, that is to say, the series, but in spite of all of our computational work, **we still do not know whether the series converges or diverges.** This complicated topic, whether a series converges or diverges, is the topic of much of this chapter. ■

We have been discussing the idea that a series is, by definition, just a sequence, the sequence of partial sums. Our next example, sheds light on this matter in a particularly clear way.

Example 7.6 *Determine whether the series*

$$\sum_{k=1}^{\infty}(-1)^k = 1 - 1 - 1 + 1 - 1 + \cdots,$$

converges or diverges.

This series may be very puzzling at first glance, and that is the whole point of the example. Do the terms cancel in pairs giving us a sum of 0?

We don't need our calculator to answer this question, but our calculator can drive home the important point. A series is a sequence of partial sums. As soon as we apply the definition, the series is no longer puzzling.

Enter $Y1 = sum(seq((-1) \wedge J, J, 1, X))$ in the Y= menu, set TblStart= 1, and ΔTbl= 1, and press [2ND] [TABLE] to see the screen shown. Does this sequence converge? Obviously not! This oscillating pattern between -1 and 0 continues forever, so that the sequence, and hence, by definition, the series itself, diverges. The point to be made here is that the practice of using the definition of a series as a sequence of partial sums can, at least in some instances, increase our understanding of a series.

Incidentally, notice what would happen if we set ΔTbl= 10. we do this frequently to scroll through sequence values faster . Now the output values in the table are all constantly -1. What does this mean? We know the sequence diverges.

7.3 Positive Term Series

The *Comparison Test (Inequality Comparison Test)* and the *Limit Comparison Test* are used to determine the convergence or divergence of a given positive term series, by comparing it to another well chosen positive term series of known convergence or divergence. The *Inequality Comparison Test* is not very suitable for calculator involvement. Using this test is more a matter of logical argument, than it is a computation. On the other hand, since our calculator can usually compute limits, the *Limit Comparison Test* can be used quite effectively. As always, our calculations do not carry the force of a proof, and limit calculations can always be forced to break down by choosing the variable close enough to its limit point.

To use a comparison test, we need a collection of positive term series of known convergence or divergence. While any positive term series, once it is known to converge or diverge, can be put into this collection, it is usually restricted, in a classroom setting, to a certain **"standard set" consisting of all of the geometric series, together with all of the so-called p-series. The well known convergent series $\sum_{n=0}^{\infty} 1/n!$ will also be placed in our standard set.**

Deciding what to compare a given series to involves "educated guess work." Since it is easy to calculate a (probable) limit value on our calculator, we can make guesses with less hesitance, knowing that it will not take much work to perform the test.

Example 7.7 *Determine whether the following series converges or diverges.*

$$\sum_{n=1}^{\infty} \frac{\sqrt{n^3+2}}{2^n}$$

Let a_n denote the nth term of the given series. The key idea here is that 2^n gets large so fast that it quickly overwhelms the numerator and makes it somewhat insignificant. This would suggest that $\sum a_n$ behaves somewhat like the series

$$\sum (\frac{1}{2})^n$$

which is a convergent geometric series. If so, then our series probably converges. Unfortunately, this geometric series is smaller than the given series and so it cannot be used in the *Comparison Test*. It is easily seen that the *Limit Comparison Test* fails for pretty much the same reason. What we need is to compare the given series to a somewhat larger convergent series. This is accomplished by using any larger geometric series which still converges. We choose to compare the given series to the series $\sum b_n$ with $b_n = \left(\frac{2}{3}\right)^n$.

Enter $Y1 = \sqrt{(X \wedge 3 + 2)/2} \wedge X$, $Y2 = (2/3) \wedge X$, and $Y3 = Y1/Y2$. The *Limit Comparison Test* prompts us to compute the limit of $Y3$ as $X \to \infty$. Deactivate $Y1$ and $Y2$, set TblStart= 1, ΔTbl= 1, and press [2ND] [TABLE] to see the values of this quotient. For a while, the values of $Y3$ are larger than 1, but eventually the values level off at a value near 0. While this is hardly a proof, it appears that

$$\lim_{n \to \infty} \frac{a_n}{b_n} = \lim_{X \to \infty} Y3(X) = L = 0.$$

This implies that the series converges. Recall that the *Limit Comparison Test* says that both series converge or both diverge if $0 < L < \infty$. In our case, $L = 0$, but the test still implies convergence. Can you see why? ■

Example 7.8 *Determine whether the following series converges or diverges.*

$$\sum_{n=1}^{\infty} \frac{(n!)^2}{(2n)!}$$

Let a_n be the nth term of the given series. The ratio test is always the most promising test to use in series involving factorials. To form the ratio $\frac{a_{n+1}}{a_n}$, press [Y=] and enter $Y1 = (2*X)!/(X!) \wedge 2$. To enter the factorial symbol, press (from the Y= menu) [MATH] ▶ ▶ ▶ to select the PRB submenu (for probability), [4] to enter the factorial symbol at the location of the cursor in the Y= menu. The ratio can now be entered as $Y2 = Y1(X+1)/Y1(X)$, or equivalently as $Y2 = Y1(X+1)/Y1$. Deactivate $Y1$, and create a table of values for $Y2$ in the usual way. The values quickly level off to a value near 0.25 as we scroll down through the table, but ERROR terms are returned for all $X \geq 34$—a disturbing event, since we want $X \to \infty$.

This example demonstrates the limits of calculator technology. We will have to try another approach, but all is not lost. Our work still suggests the "probability" that

$$\lim_{n \to \infty} \frac{a_{n+1}}{a_n} = \lim_{X \to \infty} Y2(X) = \rho \approx .25,$$

which would imply that $\rho < 1$, and this would imply convergence.

Certainly a more convincing argument is necessary before any conclusions can be drawn. It is easy to see why the error terms are generated. For $N \geq 70$, the expression $N!$, exceeds 10^{100} and generates an OVERFLOW ERROR. To avoid the problem, the expression $Y2(X)$ must be simplified by analytic means, before it is entered on our calculator. This gives us

$$\frac{a_{n+1}}{a_n} = \frac{((n+1)!)^2 (2n)!}{(2(n+1))!(n!)^2} = \frac{n^2}{(2n+1)(2n+2)}.$$

7.3. POSITIVE TERM SERIES

The simplified expression could be entered as $Y1$, so that a limit could be determined, but with this simplification, it may be apparent (without a calculator) that

$$\rho = \lim_{x \to n} \frac{a_{n+1}}{a_n} = \lim_{n \to n} \frac{n^2}{(2n+1)(2n+2)} = \frac{1}{4}.$$

Since $\rho < 1$ it follows that the series converges. ∎

What can we learn from this example? Was our calculator a useful tool? Clearly, it has its limits! In the end, we had to compute the limit analytically, but even if the conclusion had been firm, an analytic argument would have been necessary to be rigorous. Our calculator work, however, gave us an answer to anticipate, and in this sense, it was useful work. Additionally, on another example, **our calculator might suggest the probability that $\rho = 1$**. Even a rigorously determined value of **$\rho = 1$ is inconclusive**. In such a case, our preliminary calculator work might suggest that effort spent on determining analytically the value of ρ in the *Ratio Test* would probably lead to an inconclusive result. As a consequence, we could turn to alternate tests before we wasted effort on the *Ratio Test*.

Once it is known that a series converges, a partial sum can be used to approximate the value of the series. This approximation, however, is of little value unless the error between the exact sum and approximate partial sum can be estimated. For example, the series

$$\sum_{j=2}^{\infty} \left(\frac{1}{\ln(\sqrt{j})}\right)^{\ln(\sqrt{j})}$$

can be shown to converge (see Problem 12 in the exercises), but it converges so slowly that the $10{,}000$th term is still approximately 0.009 in order of magnitude. We would obviously have to add tens of thousands of terms to come close to the exact value of the series. By way of contrast, there are positive term divergent series whose partial sums get large at an almost imperceptible rate. There are many anecdotes written about how slowly the partial sums of the harmonic series get large.

> Suppose that 10 million years ago, one of our ancestors began to add up the terms of the harmonic series at the rate of 100 terms per second, and that the task was continued, without fail, each second thereafter until the present time, with the duty to continue the additions passed on smoothly from generation to generation. One can show that, at the present time, the sum of all these terms would be somewhere between 37 and 39, continuing, nevertheless, to increase, in a very slow race, to infinity. (See Problem 13 in the exercises.)

If $\sum_{k=1}^{\infty} b_k$ is a convergent series with sum s and if s_n denotes its nth partial sum, then the nth error term is defined to be

$$e(n) = \mid s - s_n \mid = \mid \sum_{k=1}^{\infty} b_k - \sum_{k=1}^{n} b_k \mid = \mid \sum_{k=n+1}^{\infty} b_k \mid.$$

Deciding how far out one needs to go in the sequence of partial sums of a convergent series in order to get a finite sum which is a good approximation for the whole series is not always an easy matter. Whenever this error term can be measured, it deserves our attention.

One of the important characteristics of the *Integral Test* is that it can be used not only to show convergence, but also to measure the size of the error between the exact sum and a partial sum. Given a series $\sum_{k=1}^{\infty} b_k$, suppose that f is a nonincreasing real valued continuous function on the interval $[1, \infty]$ such that $f(k) = b_k$ for each k. Basically the same idea used to prove the *Integral Test* can be used to prove that if the series converges then the error term satisfies

$$e(n) \leq \int_n^{\infty} f(x)dx.$$

The nonincreasing nature of the function f, in the *Integral Test* is a critical condition, but it can be relaxed. Like most tests for convergence, words like "for all" can usually be replaced by "eventually," and so it is it is only necessary to show that f is eventually nonincreasing. This complicates the error term only slightly. Using the same argument used to prove the *Integral Test* one can show that as long as f is nonincreasing for all $x \geq N$, the error term still satisfies

$$e(n) \leq \int_n^{\infty} f(x)dx \text{ for } n \geq N.$$

Example 7.9 *Show that the series $\sum_{k=1}^{\infty} \frac{k^3}{e^k}$ converges. Approximate the sum with an error not exceeding 10^{-8}.*

Based on growth patterns for the exponential function, it may be clear that the underlining function $f(x) = x^3/e^x$ is eventually decreasing. This may not, however, be clear to everyone, and we cannot use the *Integral Test* until the decreasing nature of this function is established. It is easy enough to see this, if it is not already clear, by plotting $f(x)$. More rigorously, factor $f'(x)$, to see that $f'(x) < 0$ for $x > 3$. This step is not included, and we proceed with the *Integral Test* . Since the improper integrals $\int_1^{\infty} f(x)dx$, and $\int_n^{\infty} f(x)dx$ $(n \geq 1)$ either all converge, or all diverge, there is no need to establish convergence of the series as a separate issue. If the integral involved in any error term converges, then the series converges, and so we turn our attention immediately to this error term, treating n as an unknown constant.

To compute the improper Integral, we must determine an antiderivative of $f(x)$. This process, which involves integrating by parts three times, is straight forward, but tedious to say the least. Suppose we try to avoid this lengthy computation and use a numerical technique instead. Enter $Y1 = f(X) = X \wedge 3/e \wedge (X)$ in the Y= menu. Set $Y2 = fnInt(Y1, X, 1, X)$, or if the different roles played by X in the second and fourth arguments is confusing (it's not to our calculator), then enter $Y2 = fnInt(T \wedge 3/e \wedge (T), T, 1, X)$. The two forms are equivalent. it follows that

$$e(n) \leq \int_n^{\infty} f(x)\,dx = \lim_{X \to \infty} \int_n^X f(x)\,dx = \lim_{X \to \infty} (Y2(X) - Y2(n)).$$

We construct a table of values for $Y2(X)$ to compute an approximate value for

$$A = \lim_{X \to \infty} (Y2(X)).$$

7.4. ALTERNATING SERIES AND ABSOLUTE CONVERGENCE

Deactivate $Y1$ to keep its values off of our table. Press [2ND] [TBLSET], set TblStart= 10, ΔTbl= 10, and press [2ND] [TABLE] to see the table of values. This is a demanding calculation and it takes some time to produce the screen shown. The values of $Y2$ appear to have stabilized without any need to scroll down the table. Just to be on the safe side, we go to the home screen and evaluate $Y2(200)$ to get $Y2(200) = 5.886071059$. Press [2ND] [ANS] [STO▶] [ALPHA] (A) to store the value as A. The error bound statement now takes the form $e(n) < A - Y2(n)$, where n is an unknown.

The end is near. We wish to determine an n for which $e(n) < A - Y2(n) < 10^{-8}$. Enter $Y3 = A - Y2 - 10 \wedge (-8)$. There are several ways to proceed. We create a table of values. Deactivate $Y2$, and set up a table with TblStart= 10, and ΔTbl= 10. It will take some time to generate the table. Once the table appears, we can see that $Y3(20)$ is positive and $Y3(30)$ is negative (along with all subsequent values). It follows that the desired value of n satisfies $20 < n < 30$. Reset TblStart=20 and ΔTbl= 1, press [2ND] [TABLE] (another wait—relax, enjoy the break), and scroll down to see the table shown, where $Y3(28) = 6.9\,10^{-9}$ and $Y3(29) = -3\,10^{-9}$. It follows that $n = 29$. We compute $sum(seq(Y1, X, 1, 29))$ to get

$$\sum_{k=1}^{\infty} \frac{k^3}{e^k} \approx \sum_{k=1}^{29} \frac{k^3}{e^k} = 6.00065127,$$

which has 8 decimal places of accuracy.

A more rigorous approach to this problem might involve an analytic computation of the exact antiderivative $F(x)$ of $f(x)$. This involves integrating by parts three times, sending $\int x^3 e^{-x}\,dx \to \int x^2 e^{-x}\,dx \to \int x e^{-x}\,dx \to \int e^{-x}\,dx$ with the last integral easily evaluated. The details are omitted. It turns out that

$$F(x) = -\frac{x^3 + 3x^2 + 6x + 6}{e^x}.$$

The function $F(X)$ can then be used as a more reliable version of $Y2$. ∎

Interestingly enough, our approximate sum of $s = 6.0065127$ is correct to this many decimal places, even though we had to sum very few terms in the series to reach this degree of accuracy.

7.4 Alternating Series and Absolute Convergence

The error term can also be estimated quite easily for an alternating series. According to the *Alternating Series Test*, if $\{a_k\}$ is a nonincreasing sequence of positive numbers which converges to 0, then the alternating series $\sum_{k=1}^{\infty}(-1)^{k+1}a_k$ converges, and for each n, the error term satisfies

$$e(n) \leq a_{n+1}.$$

If the sequence $\{a_k\}$ is nonincreasing only for $k \geq N$, then it is easy to see that the error term still satisfies

$$e(n) \leq a_{n+1} \text{ for } n \geq N.$$

Example 7.10 *Show that the sequence $\sum_{k=1}^{\infty}(-1)^k \frac{1}{k^2+2^k}$ converges and approximate its sum with an error not exceeding 10^{-8}.*

This series clearly converges by the *Alternating Series Test*, and so we consider the error term immediately. Set $Y1 = 1/(X \wedge 2 + 2 \wedge X)$. Notice that the alternating sign $(-1)^k$ is not a part of this expression. Press [2ND] [TBLSET] and set TblStart= 1 and ΔTbl= 10. Press [2ND] [TABLE] to see the values of $Y1$. Observe that $Y1(21) = 4.8 \, 10^{-7}$ and $Y1(31) = 5 \, 10^{-10}$. Consequently, we reset TblStart= 21 and ΔTbl= 1. Press [2ND] [TABLE] to get the screen shown. For $n = 26$, we conclude that $e(n) \leq a_{n+1} = a_{27}$ has the required degree of smallness. From the home screen, enter $sum(seq((-1) \wedge (X) * Y1, X, 1, 26)$, and press enter to get

$$\sum_{k=1}^{\infty}(-1)^k \frac{1}{k^2+2^k} \approx \sum_{k=1}^{26}(-1)^k \frac{1}{k^2+2^k} = -0.2470767613.$$

We should drop the last two digits in this answer, because of the size of the error term, and this gives us an approximate sum of $s = -0.24707676$. This much accuracy is quite striking considering that we only added the first 27 terms of the series to obtain this approximation.

Example 7.11 *Determine whether the series $\sum_{n=1}^{\infty}(-1)^n n^{-1+1/n}$ converge absolutely, conditionally, or diverges.*

When problems like this are solved by hand, the series under consideration is usually studied with some care before a decision is made regarding which convergence test to use. The labor involved in using these tests is often great enough that there is a reluctance to use a test without a reasonable expectation of a successful outcome. With our calculator, on the other hand, the computations are no cause for concern, and so there is no real need to be so cautious before a test is used. **Even if a more rigorous analytic argument is desired, a calculator can help us decide in advance which test is likely to lead to a decisive result.**

Let $a_n = n^{-1+1/n}$. We first consider the absolute value series, and compare it to the $p = 2$ series $\sum b_n$, $(b_n = \frac{1}{n^2})$. Enter $Y1 = X \wedge (-1 + 1/X)$, $Y2 = 1/X \wedge 2$, $Y3 = Y1/Y2$. It is readily seen that table values for $Y3$ get large suggesting that

$$\lim_{n \to \infty} \frac{a_n}{b_n} = \infty.$$

7.4. ALTERNATING SERIES AND ABSOLUTE CONVERGENCE

With such a limit value, the comparison test would be inconclusive. **This is not, however, a wasted effort, and a failed test often suggests a new direction.** Since $a_n/b_n \to \infty$ as $n \to \infty$, it must mean that $b_n \to 0$ faster than $a_n \to 0$ as $n \to \infty$. To produce a finite limit, we must slow down the rate at which $b_n \to 0$. It seems reasonable, then, to use $b_n = 1/n$ instead, and to compare $\sum a_n$ to the divergent Harmonic Series $\sum \frac{1}{n}$. It is easy to simply edit $Y2$ to form $Y2 = 1/X$. With this change, a table of values for $Y3$ suggests that

$$\lim_{n \to \infty} \frac{a_n}{b_n} = L = 1.$$

With such a value for L, the test is conclusive, and it follows that the series $\sum a_n$ diverges (since the Harmonic Series diverges), so the original series does not converge absolutely.

According to the *Alternating Series Test*, the series will converge if the sequence $\{a_n\}$ ($a_n = n^{-1+1/n}$) is a decreasing sequence which converges to 0. The last limit computed above clearly implies that the sequence converges to 0 (and then some). Showing that the sequence decreases can be done with a plot, or a Table of values. It can also be done somewhat more rigorously by showing that its derivative is always negative. The expression $f(x) = x^{-1+1/x}$ can be differentiated logarithmically, or it can be converted to an expression of the form $f(x) = e^{g(x)}$, and then differentiated. Either way, the derivative

$$f'(x) = \left(\frac{1 - \ln(x) - x}{x^2} \right) x^{-1+1/n}.$$

is clearly always negative, so $f(x)$ and hence $\{a_n\}$ are decreasing expressions. By the *Alternating Series Test*, the given series $\sum (-1)^n a_n$ converges. Combining both results, it follows that the given series converges conditionally. ∎

Example 7.12 *Determine whether the series $\sum_{n=1}^{\infty} (-1)^n \left(\frac{n^{\ln(n)}}{2\sqrt{n}} \right)$ converge absolutely, conditionally, or diverges.*

Let $a_n = \frac{n^{\ln(n)}}{2\sqrt{n}}$. We try the *Root Test* first. Set $Y1 = X \wedge \ln(X)/2 \wedge \sqrt{(X)}$, and $Y2 = (Y1) \wedge (1/X)$. Deactivate $Y1$, set TblStart= 1, ΔTbl= 10, and press [2ND] [TABLE] to see the table. Scroll down through the values to see that they seem to converge to a value of 1. The convergence is slow, however, so to confirm are suspicions, we should increase the value on ΔTbl, or use some other technique to scroll more quickly through table values. A value of ΔTbl= 100 was used to produce the screen shown. The evidence is fairly convincing that the Root Test value of $\rho = \lim_{\to \infty} \sqrt[n]{a_n}$ is $\rho = 1$. This makes the *Root Test* inconclusive.

X	Y2
1212	1.0219
1312	1.0204
1412	1.019
1512	1.0178
1612	1.0167
1712	1.0158
1812	1.0149

X=1812

Next we try a comparison with the $p = 2$ series $\sum b_n$ ($b_n = \frac{1}{n^2}$). **The experience is very disturbing, and it presents us with a dilemma for which there is no easy answer.** We leave $Y1$ as it is currently entered, and enter $Y2 = 1/X \wedge 2$ and $Y3 = Y1/Y2$. Deactivate $Y1$ and $Y2$, set TblStart= 1 and ΔTbl= 10. Press [2ND] [TABLE] to see the table and scroll down through table values to look for a limiting value. Evidently, $Y3 \to \infty$ as $X \to \infty$. To confirm our suspicions, increase the value of ΔTbl to create the table shown.

X	Y3
607	9.7E15
707	2.5E16
807	5.2E16
907	9.8E16
1007	1.7E17
1107	2.6E17
1207	3.7E17

X=1207

It would seem reasonable to conclude that $\lim_{X \to \infty} Y3(X) = \infty$, but that conclusion would be FALSE! We simply have not scrolled far enough down the table. To speed through table values at a faster rate, enter $Y4 = 2 \wedge X$, and $Y5 = Y3(Y4)$. We have used this technique before. It creates a table of values of the form $Y3(2^X)$ ($X = 1, 2, \ldots$). The values of 2^X increase slowly at first, but rapidly as X grows, with the rate of increase far outpacing what we would achieve by setting ΔTbl equal to a large number. Deactivate $Y3$, set TblStart= 1, ΔTbl= 1, and press [2ND] [TABLE] to see the table. We don't have to scroll very far to see the screen shown, where the values of $Y5$ have finally turned around and are decreasing to 0. This suggests that

X	Y4	Y5
12	4096	1E18
13	8192	7E15
14	16384	6.2E10
15	32768	30.67
16	65536	1E-14
17	131072	ERROR
18	262144	ERROR

X=16

$$\lim_{n \to \infty} \frac{a_n}{b_n} = \lim_{X \to \infty} Y5(X) = L = 0,$$

which would imply that the given series $\sum (-1)^n a_n$ converges absolutely.

It looks like $Y5$ decreases rapidly to 0, but as the table shows, this is only because the argument $Y4$ of $Y5$ is increasing so rapidly. It is also evident that we have allowed $Y4$ to become so large that OVERFLOW errors were generated. While the evidence in support of absolute convergence is strong, the error terms certainly make our conclusions less convincing. Shall we go on? The limit of a_n/b_n could be computed analytically by using *L'Hopital's Rule*, but before the rule could be used, the expression would have to be written in a form more suitable for differentiation. Ultimately, this would lead us to take the logarithm of the expression a_n/b_n, and this, by the way, suggests another calculator approach. Write

$$c_n = \ln\left(\frac{a_n}{b_n}\right) = \ln\left(\frac{n^2 \, n^{\ln(n)}}{2^n}\right) = (2 + \ln(n))\ln(n) - n\ln(2).$$

Look at the simplified expression for c_n and you might be able to see, without a calculator, that $c_n \to -\infty$ as $n \to \infty$. This is because **n grows large faster than $\ln(n)$ or $(\ln(n))^2$**. Even without this kind of reasoning, c_n is certainly in a form which can be reliably entered on our calculator, and then it easily follows that $\lim_{n \to \infty} c_n = -\infty$. Consequently,

$$L = \lim_{n \to \infty} \frac{a_n}{b_n} = \lim_{n \to \infty} e^{c_n} = e^{-\infty} = 0.$$

7.5. POWER SERIES

Now we can say with more confidence that the given series converges absolutely. ∎

This example demonstrates quite dramatically the **problem involved in using our calculator to compute a limit**. Unless our result is backed up by an analytic argument, there will **always be a potential for being mislead by the data**, as we were with our first attempt to compute the limit of a_n/b_n. This risk can be minimized, but never entirely eliminated, by keeping an open mind about the evaluation process, and by looking at tables in a variety of ways. We may always want to consider some table work with $\Delta\text{Tbl}=1$. (Recall the warning following Example 7.6, where we generated misleading data by letting $\Delta\text{Tbl}=10$.) We should, however, also look far down the table by changing the value of ΔTbl from 1 to 10, 100, and perhaps 1000 and larger in order to scroll down the table at a faster rate. Finally, replacing X by 2^X, allows us to scroll down through table values at an even faster rate. At the other extreme, we run the risk of quickly generating OVERFLOW errors, if we choose numbers that get too large too quickly.

7.5 Power Series

The radius of convergence is found by using the *Ratio* or *Root Test*. These tests can only be used on positive term series, or, in other words, to test for absolute convergence. View a power series $\sum a_n(x-x_0)^n$ as $\sum p_n$, where $p_n = a_n(x-x_0)^n$. The *Ratio Test* applied to the series $\sum p_n$ gives

$$\rho = \lim_{n\to\infty}\left|\frac{p_{n+1}}{p_n}\right| = \lim_{n\to\infty}\left|\frac{a_{n+1}(x-x_0)^{n+1}}{a_n(x-x_0)^n}\right| = \left(\lim_{n\to\infty}\left|\frac{a_{n+1}}{a_n}\right|\right)|x-x_0| = R|x-x_0|,$$

where

$$R = \lim_{n\to\infty}\left|\frac{a_{n+1}}{a_n}\right|.$$

Determining where the power series converges absolutely and where it diverges comes from an evaluation of R.

Example 7.13 *Find the exact interval of convergence of the power series*

$$\sum_{n=1}^{\infty}\frac{n^{\sqrt{n}}}{2^n}x^n.$$

We could try a direct evaluation of R by entering $Y1 = X \wedge \sqrt{(X)}/2 \wedge X$ and $Y2 = Y1(X+1)/Y1(X)$. However, $Y2$ converges very slowly towards its limit, and before sufficient accuracy is obtained, OVERFLOW errors are experienced because of the large terms involved in the various parts of $Y1$. The simplification

$$\frac{a_{n+1}}{a_n} = \frac{2^n(n+1)^{\sqrt{n+1}}}{2^{n+1}n^{\sqrt{n}}} = \frac{(n+1)^{\sqrt{n+1}}}{2\,n^{\sqrt{n}}}$$

helps somewhat, but the terms are still large enough to cause OVERFLOW problems. We compute instead the limit of

$$c_n = \ln\left(\frac{a_{n+1}}{a_n}\right) = \ln\left(\frac{(n+1)^{\sqrt{n+1}}}{2n^{\sqrt{n}}}\right) = \sqrt{n+1}\ln(n+1) - \sqrt{n}\ln(n) - \ln(2).$$

This strategy has been used before to bring down the magnitude of our computations.

Enter $Y1 = \sqrt{(X)} * \ln(X)$, $Y2 = Y1(X+1) - Y1 - \ln(2)$, deactivate $Y1$ and create a table of values for $Y2$ in the usual way. This seems to suggest that $Y2$ is converging to some value, but the convergence is very slow. Fortunately, $Y2$ can be evaluated when X is exceedingly large (our logarithmic reduction of values has helped). After sampling values of $Y2$ for large values of X to get a reliable determination, we go to the home screen, enter $Y2(10 \wedge 20)$ and press [STO▶] [ALPHA] ([A]) [ENTER] to store our estimated limit value as $A = -.6931471806$. It follows that

$$R = \lim_{n \to \infty} \frac{a_{n+1}}{a_n} = \lim_{X \to \infty} Y1(X) = \lim_{X \to \infty} e^{Y2(X)} = e^A = 0.5 = \frac{1}{2}.$$

The series converges for $|x| < 2$ (where $\rho = \frac{1}{2}|x| < 1$), and diverges for $|x| > 2$ (where $\rho > 1$). It clearly diverges at $x = \pm 2$, and so the exact interval of convergence is the open interval $(-2, 2)$.

Example 7.14 *A function $f(x)$ is defined by the power series*

$$f(x) = \sum_{n=2}^{\infty} \frac{[(-1)^n \ln(n)]^2}{n}(x-3)^n.$$

The domain of f is the interval of convergence, and the sequence $\{p_n\}$ of partial sums of the series is a sequence of polynomials in x which approximate f. Determine the domain of f and approximate the graph of f by graphing the polynomial $p_{20} = p_{20}(x)$

We use the *Ratio Test* to establish the interval of convergence in this example. Let $a_n = \frac{(-1)^n [\ln(n)]^2}{n}$. One would not expect a problem in the evaluation of $|a_n|$, and so we enter $Y1 = \ln(X) \wedge 2/X$ (without the alternating negative sign), $Y2 = Y1(X+1)/Y1$, and create a table of values for $Y2$ in the usual way. There are no complications, and we conclude that

$$R = \lim_{n \to \infty} \left|\frac{a_{n+1}}{a_n}\right| = \lim_{X \to \infty} Y2(X) = 1.$$

It follows that the series converges for $\rho = R|x-3| = |x-3| < 1$, and diverges for $\rho = |x-3| > 1$. At $x = 2$, the series is the positive term series $\sum |a_n|$ which is clearly larger than the harmonic series, and so it diverges. At $x = 4$, the series is the alternating series $\sum a_n$, which will converge by the *Alternating Series Test*, if we can show that $\{|a_n|\}$ is a decreasing sequence which converges to zero. We activate $Y1$ and deactivate $Y2$. The result is established in a straight forward manner either analytically, or by looking at a plot or a table of values of $Y1$. It follows that the series converges, and so the exact interval of convergence is the half open interval $(2, 4]$.

To finish the example, we approximate the graph of f over the interval $(2,4]$ by graphing the polynomial

$$p_{30}(x) = \sum_{n=2}^{30} \frac{(-1)^n [\ln(n)]^2}{n}(x-3)^n.$$

7.5. POWER SERIES

Deactivate $Y1$ and $Y2$ and enter

$Y3 = sum(seq((-1) \wedge N * Y1(N) * (X - 3) \wedge N, N, 2, 30)).$

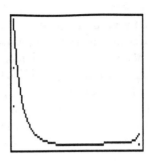

The series diverges at $x = 2$, so it may be a good idea to let X_{min} be a number slightly larger than 2. Press [WINDOW] and enter $X_{min} = 2.1$ and $X_{max} = 4$. Press [ZOOM] [0] (which selects ZoomFit) to see the screen shown. Start the graph just before you take a break for dinner. It should be done by the time you are back from your break. ■

If f is a function which has derivatives of order $j = 1, 2, \ldots n$ in some open interval containing $x = a$, then the *Taylor Series* of order n based at $x = a$ is defined to be

$$P_n(x) = f(a) + f'(a)(x - a) + \frac{f''(a)}{2!}(x - a)^2 + \cdots + \frac{f^{(n)}(a)}{n!}(x - a)^n.$$

Example 7.15 *Find the Taylor series for $f(x) = (2 - x)e^x$ based at $x = 0$, and let $p_n(x)$ denote its nth partial sum (i.e., the Taylor polynomial of order n for $f(x)$ based at $x = 0$). Plot $f(x)$ and $p_{10}(x)$ on the interval $[-2, 2]$. Plot the error term $E_{10}(x) = f(x) - p_{10}(x)$ on the same interval. Evaluate $f(5.8)$ and $p_n(5.8)$ for $n = 10, 20, 30$.*

We could determine the Taylor series for $f(x)$ directly by computing $f^{(n)}(0)$ for each $n = 1, 2, \ldots$, but it is easier to compute its Taylor series indirectly from the well known Taylor series for e^x. We write

$$f(x) = (2 - x)e^x = (2 - x) \sum_{k=0}^{\infty} \frac{x^k}{k!} = 2 \sum_{k=0}^{\infty} \frac{x^k}{k!} - x \sum_{k=0}^{\infty} \frac{x^k}{k!} = \sum_{k=0}^{\infty} 2\frac{x^k}{k!} - \sum_{k=0}^{\infty} \frac{x^{k+1}}{k!}.$$

In the second sum, if we use k instead of $k+1$ for the power, then the corresponding factorial will be one less than k or $(k-1)!$, so that

$$\sum_{k=0}^{\infty} \frac{x^{k+1}}{k!} = \sum_{k=1}^{\infty} \frac{x^k}{(k-1)!}.$$

This is done so that the both sums have a common term of x^k. The first term in the first sum must be pulled out, so that both sums start out at $k = 1$, then we have

$$f(x) = 2 + \sum_{k=1}^{\infty} 2\frac{x^k}{k!} - \sum_{k=1}^{\infty} \frac{x^k}{(k-1)!} = 2 + \sum_{k=1}^{\infty} \left(\frac{2}{k!} - \frac{1}{(k-1)!} \right) x^k,$$

which gives us the desired Taylor series.

We can now enter the Taylor polynomial in the Y= menu. This takes a great deal of keyboard activity, and to avoid doing it more than once, we let N denote the upper limit on the sum. It will be an easy matter to set $N = 10$, and then to change the value of N from $N = 10$ to $N = 20, 30$. We enter

$Y1 = 2 + sum(seq((2/K! - 1/(K - 1)!) * X \wedge K, K, 1, N)), Y2 = (2 - X) * e \wedge (X).$

Any graph involving $Y1$ will take a great deal of computational time to generate. Recall that the (!) operator is available in the PRB submenu of the MATH menu. We go to the home screen, enter 10 and press [STO▶] [ALPHA] [N] [ENTER], to store the value $N = 10$. Set $X_{min} = -2$, $X_{max} = 2$, and press [ZOOM] [0] (which selects ZoomFit) to get the screen shown. Only one graph seems to appear, because the two graphs are indistinguishable. This is dramatic evidence concerning how well $p_{10}(x)$ approximates $f(x)$, but there is a much better way to demonstrate graphically how well $p_{10}(x)$ approximates $f(x)$. By graphing

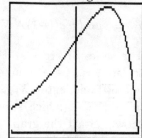

$$E(x) = p_{10}(x) - f(x)$$

under ZoomFit, we will not only be able to see the error term, but the scale used on the vertical axis will be small enough that we will be able to see even small values of the error term. Enter $Y3 = Y1 - Y2$, deactivate $Y1$ and $Y2$ and graph $Y3$ under ZoomFit to get the screen shown. The small vertical scale does not show on this screen. Press [WINDOW] and notice that $Y_{min} = -3.887 \, 10^{-4}$ and $Y_{max} = 5.644 \, 10^{-4}$. This small range implies that our approximation is quite accurate even though the order $N = 10$ of the Taylor polynomial $p_{10}(x)$ is quite small.

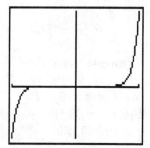

We still have a few evaluations to compute. Enter $Y2(5.8)$ and press [ENTER] to get $f(5.8) = -1255.138328$. In the same way, enter $Y1(5.8)$ to get $p_{10}(5.8) = -494.6047284$—not a very good approximation. Because we used a letter N instead of 10 when we set up the function $Y1$, it is now an easy matter to change the number of terms in the sum with very little keyboard activity. Enter 20 and press [STO▶] [ALPHA] [N] [ENTER] to set $N = 20$. Press [2ND] [[ENTRY]] [2ND] [[ENTRY]] to bring $Y1(5.8)$ back to the active line, and press [ENTER] to get $p_{20}(5.8) = -1255.132822$—much better. Enter 30 and press [STO▶] [ALPHA] [N] [ENTER] to set $N = 30$. Press [2ND] [[ENTRY]] [2ND] [[ENTRY]] to again bring $Y1(5.8)$ back to the active line, and press [ENTER] to get $p_{30}(5.8) = -1255.138328$—an impressive approximation for $f(5.8)$. ∎

By looking at this graph, we can see, visually at least, how close a Taylor polynomial is to the graph of the function f that it represents. To get a more reliable estimate of the error term involved, we consider the **remainder term** in *Taylor's Theorem*. If $p_n(x)$ is the Taylor polynomial of order n for f based at $x = x_0$, then the error term is defined by

$$E_n(x) = |\, f(x) - p_n(x) \,|.$$

If f has $(n+1)$ derivatives on an interval I centered at $x = x_0$, then *Taylor's Theorem* says that for x in I,

$$E_n(x) = \frac{|f^{(n+1)}(z)|}{(n+1)!}|x - x_0|^{n+1},$$

7.5. POWER SERIES

where z is some point between x_0 and x. Since we generally are not able to find z, we try to get rid of it. If M is the maximum value of $|f^{(n+1)}(z)|$ for z in I, we usually use instead the error bound

$$E_n(x) = \frac{|f^{(n+1)}(z)|}{(n+1)!}|x - x_0|^{n+1} < \frac{M}{(n+1)!}|x - x_0|^{n+1}.$$

As a practical issue, there is usually no advantage gained in computing M accurately, since clearly any number larger than M can be used in place of M. Using a number slightly larger (but simpler) than M would generate a slightly weaker error term, but the effect is usually so slight that it is of no consequence. Consequently, the usual practice is to use some easily obtained number known to be larger than every value of $|f^{(n+1))}(z)|$ for z in I.

In the next example, notice how a value for M is easily selected, visually, off of a graph.

Example 7.16 *Let $f(x) = e^x \cos(x)$, and let $p_{16}(x)$ be the Taylor polynomial of order 16 for $f(x)$ based at $x_0 = 0$. Use the Taylor Theorem to determine an upper bound on the error $E_{16}(x)$ for x in the interval $[-3, 3]$.*

We need to compute the 17th derivative $f^{(17)}(x)$. That may seem like a daunting task, but because of the regularity of the function f, there are some interesting symbolic maneuvers that can minimize the task. Let g be the "parallel" function $g(x) = e^x \sin(x)$. It is easy to establish from $f'(x) = e^x \cos(x) - e^x \sin(x)$, that $f' = f - g$. In the same way, $g' = f + g$. From these two formulas we can move forward quickly. It follows that $f'' = f' - g' = (f - g) - (f + g) = -2g$. In the same way $g'' = 2f$. Consequently, $f^{(4)} = -2g'' = -2(2f) = -4f$, so that $f^{(8)} = -4f^{(4)} = -4(-4f) = 16f$. Thus $f^{(16)} = 16f^{(8)} = 16(16f) = 2^8 f$, and $f^{(17)} = 2^{16} f' = 2^{16}(f - g)$.

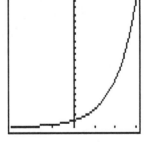

We find the largest value M of $f^{(17)}(z)$ from a graph. Enter $Y1 = e \wedge (X) \cos(X)$, $Y2 = e \wedge (X) \sin(X)$, $Y3 = abs(Y1 - Y2)$, and deactivate $Y1$ and $Y2$. The absolute value function abs( ) is in the NUM submenu of the MATH menu. Plot $Y3$ under ZoomFit with $X_{min} = -3$ and $X_{max} = 3$ to get the screen shown. Press [TRACE] and move the trace cursor to the high point of $Y = 22.7$ at $X = 3$. Using an upper bound of 23 for $f - g$, we use $M = 2^{16} 23$ as an upper bound for $f^{(17)}$. Therefore, an error bound for x in the interval $I = [-3, 3]$ is

$$E_{16}(x) < \frac{2^{16} 23}{17!}|x|^{17} = 4.23778824 \, 10^{-9}|x|^{17} < 4.3 \, 10^{-9}|x|^{17}.$$

For example, the error terms generated by approximating $f(x)$ by $p_{16}(x)$ for $x = 0.8$ and $x = 1.4$ are

$$E_{16}(0.8) < 4.3 \, 10^{-9} (0.8)^{17} < 9.7 \, 10^{-11}, \quad E_{16}(1.4) < 4.3 \, 10^{-9} (1.4)^{17} < 1.4 \, 10^{-6}$$

7.6 Exercise Set

1. A sequence $\{a_n\}$ is defined by each of the following formulas. In each case list a representative sample of its terms as decimal numbers. Describe the behavior of the sequence. If it converges, compute its limit analytically.
    a) $a_n = \frac{n}{2^n}$ for $n = 1, 2, \ldots$ b) $a_n = \frac{n^2}{10n+100}$ for $n = 1, 2, \ldots$

2. A sequence $\{a_n\}$ is defined by each of the following formulas. In each case list a representative sample of its first 50 terms as decimal numbers. By looking at this list, does the sequence appear to converge or diverge? If it appears to converge, try to predict its limit? Compute a_{400} as a decimal. Does this confirm or contradict your previous conclusions. If it converges, use your calculator to determine the limit L of the sequence as a decimal. (Strive to use your calculator in a convincing way.) If the sequence converges, or diverges to $L = \pm\infty$), determine an integer n, and a decimal value a_n such that a_n is "close" to the limit of the sequence, or has order of magnitude at least as large as 10^{10} if $L = \pm\infty$. (You may have to go quite far out in the sequence to find such a term.)
    a) $a_n = (n + (-1)^n \sqrt{n}) \sin(\frac{1}{n})$ for $n = 1, 2, \ldots$
    b) $a_n = \frac{(\ln(n))^{\ln(n)}}{n^5}$ for $n = 2, 3, \ldots$

3. A sequence $\{a_n\}$ is defined by each of the following formulas. In each case, use your calculator in a convincing way to determine the limit of the sequence, if it converges, or state that the sequence diverges. Give some sample values to support your claim. Some of the sequences may have obvious limits which can be evaluated without a calculator, by a simple observation. Identify such sequences, and determine their limits, before using a calculator. If the sequence does not have an obvious limit, identify its indeterminate form before you attempt to compute its limit. **Be skeptical of your calculator's performance**. If necessary, devise an alternate way to express a_n as a decimal, in such a way that the decimal value of a_n gets close to the limit of the sequence, as n gets large.
    a) $a_n = \sec(\frac{1}{n})^{5n^2}$ for $n = 1, 2, \ldots$
    b) $a_n = (1 + \tanh(n))^{1/n}$ for $n = 1, 2, \ldots$
    c) $a_n = (1 - \tanh(n))^{1/n}$ for $n = 1, 2, \ldots$
    d) $a_n = \frac{n^{2n+1}}{(n+1)^n}$ for $n = 1, 2, \ldots$

4. The sequence $\{a_n\}$ in problem 4c is quite interesting. Explain (in a very specific numerical way) why your calculator gives such a false table of values. Its behavior here is clear and predictable, and in order to understand the limitations of technology, it is important to pursue this question. Determine the limit of this sequence analytically.

5. Let L denote the limit of the sequence $\{a_n\}$ in Problem 4c. Find an N such that $|a_n - L| < 10^{-3}$ for all $n > N$. This turns out to be a nontrivial problem. First try to find N with a graphical approach. If this fails to be useful, try to find N by experimenting numerically with different values of n. Why do these methods fail so dramatically? How can this failure be avoided? One approach that works involves converting the hyperbolic tangent to its exponential form. Once you find

7.6. EXERCISE SET

an integer n for which a_n is sufficiently close, how do you know that all subsequent terms in the sequence will also be sufficiently close? It may help to establish that the sequence $\{a_n\}$ (in its converted form) is either increasing or decreasing.

6. Let $\{x_n\}$ be the sequence defined recursively by $x_n = \frac{n^2+5}{n^2} x_{n-1}$ for $n = 2, 3, \ldots$ with $x_1 = 3$. List a representative sample of its terms as decimal numbers. Describe the behavior of the sequence. Use your calculator to determine its limit, or to show that the sequence diverges.

7. Let $\{b_n\}$ be the sequence defined recursively by $b_n = 2b_{n-1} + 3b_{n-2}$, $b_1 = 1, b_2 = 2$. List a representative sample of its terms as decimal numbers. Use your calculator to determine its limit, or to show that the sequence diverges.

8. Suppose that one month before the first day of your college education, your parents put $70,000 in an account earning 8% annual interest compounded monthly. You are allowed to draw out $1300 per month from this account for living and educational expenses, with your first draw occurring one month after this lump sum deposit. What ever is left after you graduate is a graduation gift. Determine the sequence $\{b_n\}$ of monthly balances. If the sequence is defined recursively, then no complicated finance formulas are needed. If you graduate in 48 months, what is the value of your graduation gift. How long can you go to school before the money runs out.

9. A low level salt contamination is discovered in a water reservoir containing 80 million gallons of water—roughly a one year supply of water for a small community. The problem is fixed, and pure water begins to flow into the reservoir again. Each month a volume of pure water equal to that month's usage is pumped into the reservoir. It mixes (instantly) with all of the contaminated water in the reservoir. The contaminated water is then pumped out of the reservoir for use in the community, which is reluctantly willing to use the contaminated water rather than throw all 80 million gallons of it away. Over time the contamination level is gradually reduced, but how long will this take, and how long will the contamination remain at an undesirable level? To answer such questions, we need to determine the sequence $\{b_j\}$ of monthly contamination levels in units of mg/gal (milligrams of salt per gallon of water). In January (month $j = 0$ of the study), the contamination level was $b_0 = 23\, mg/gal$. Water usage naturally oscillates during the year with less water used in January than in July. A formula for monthly water usage is determined to be $w_j = 6\,10^6(1 - 0.12\cos(\pi j/6))$ gallons of water. Notice that the low points of $w_0 = w_{12} = w_{24} \ldots = 5,280,000$ gallons of water correspond to January, and then January again of all the subsequent years, while the high points of $w_6 = w_{18} = w_{30} = \ldots = 6,720,000$ gallons of water correspond to July, and then July again of subsequent years.

Determine the sequence of monthly contamination levels. (This may appear to be a very daunting task, but it is a very straight forward problem **if the sequence is defined recursively**.) How many months will it take for the contamination level to be reduced by half? How long will it take for the level to be reduced to $5\, mg/gal$, and how long to be reduced to $1\, mg/gal$? Knowing that the community draws

72,000,000 gallons of water a year from the reservoir may help you judge whether your answers seem reasonable or not.

10. Suppose that an annuity earning 7.2% interest compounded monthly is set up with an initial deposit of $2,000. The plan calls for monthly deposits starting at $400. Each month the deposits are to be increased by $\frac{2}{5}$%. (In practice, the payments would actually be held constant over each calendar year and an equivalent increase in monthly deposits would be put into effect each January. The monthly increase, however, makes the problem easier to model.) What is the value of the annuity after 30 years?

11. Some of the mystery surrounding the notion of a series can be removed by appreciating that a series is just a sequence—the sequence of partial sums. Viewing a series in this way may seem unnecessary in calculations, but to fully understand what a series is, as a concept, it helps to pay some attention to the definition of a series. With this in mind, use your calculator to set up the sequence of partial sums of the series $\sum_{j=0}^{\infty} \frac{1}{n!}$. List the first 20 terms of this sequence. (The output should be a sequence of decimal numbers.) You will soon know, if you do not already know, that this series converges to e. Compare the 20 terms in this sequence to a decimal expansion for e.

12. Determine whether the following series converge absolutely, conditionally, or diverge (p denotes a fixed real number).

   a) $\sum_{j=1}^{\infty} \frac{j^4}{1.2^j}$ b) $\sum_{j=2}^{\infty} \frac{100^{2j}}{j!}$
   c) $\sum_{j=2}^{\infty} (1 - \sqrt[j]{j})^j$ d) $\sum_{j=2}^{\infty} (\sqrt{4 + j^2} - j)$
   e) $\sum_{j=2}^{\infty} (\frac{1}{\ln(j)})^{\ln(j)}$ f) $\sum_{j=2}^{\infty} (\frac{1}{\ln(\sqrt{j})})^{\ln(\sqrt{j})}$
   g) $\sum_{j=3}^{\infty} (\frac{1}{\ln(\ln(j))})^{\ln(\ln(j))}$ h) $\sum_{j=2}^{\infty} (\frac{1}{j \ln(j)})$
   i) $\sum_{j=2}^{\infty} (\frac{1}{j(\ln(j))^2})$ j) $\sum_{j=2}^{\infty} (\frac{1}{\ln(j)})^p$
   k) $\sum_{j=1}^{\infty} \frac{(j+1)^j}{j^j j!}$ l) $\sum_{j=1}^{\infty} \frac{(j!)^3}{j\sqrt{j}}$

13. The inequality used to prove the *Integral Test* can be used to prove the inequality

$$\ln(n+1) < \sum_{j=1}^{n} \frac{1}{j} < 1 + \ln(n).$$

Use this inequality to establish the claim made in the anecdote about the harmonic series on page 115.

14. Show that the series $\sum_{k=2}^{\infty} \frac{1}{k(\ln(k))^6}$ converges. Determine an n such that the error term $e(n) < 10^{-4}$. Use this to approximate the sum of the series with this much accuracy.

15. Show that the series $\sum_{k=0}^{\infty} \frac{(-4)^k}{k!}$ converges. Determine an n such that the error term $e(n) < 10^{-8}$. Use this to approximate the sum of the series with this much accuracy.

16. Find the exact interval of convergence of each of the power series.
   a) $\sum_{n=2}^{\infty} \frac{\ln(n)}{n^2 4^n} x^n$ b) $\sum_{n=2}^{\infty} \ln(n)^{\ln(n)} (x+3)^n$

7.6. EXERCISE SET

17. For each of the following, compute the Taylor polynomial of order 8 based at the given point a. Compare each Taylor polynomial with the function it represents, by plotting both of them on the same graph over an appropriate interval.
    a) $\cos(x)$, $a = 0$ b) e^x, $a = 0$ c) $\ln(x)$, $a = 1$ d) $\sqrt{x}$, $a = 9$

18. Let $f(x) = x\cos(\sqrt{x})$. Determine the Taylor polynomial of order $n = 3$ for $f(x)$ based at $x_0 = 50$. Plot $f(x)$ along with its Taylor polynomial on the interval $30 \le x \le 70$. Use Taylor's remainder formula to estimate the error involved in estimating $f(x)$ by its Taylor polynomial. Compare this estimate to the actual difference between $f(x)$ and its Taylor polynomial by plotting this difference function on the interval $30 \le x \le 70$.

19. Let $f(x) = \cos(x)\cosh(x)$, Let $p_{18}(x)$ be its Taylor series of order 18 based at $x_0 = 0$. Use Taylor's Theorem to determine an upper bound on the error $E_{18}(x) = |f(x) - p_{18}|$ in terms of x.

    The pattern of derivatives $f^{(j)}(x)$ can be determined in a manner similar to our work in Example 7.16. The function f has three other "parallel" functions: $f_1(x) = \cos(x)\sinh(x)$, $f_2(x) = \sin(x)\cosh(x)$, $f_3(x) = \sin(x)\sinh(x)$.

An application of Taylor Series

When the antiderivative of a continuous function $f(x)$ cannot be found analytically, we can still specify the antiderivative by using the *Fundamental Theorem of Calculus*. We have used this approach on several previous occasions. Our calculator is well suited to compute numerical values for the antiderivative $Y2 = fnInt(f(X), X, A, X)$ of $f(X)$, where A is a constant, **but we can only do this once.** We cannot use our calculator to compute a numerical antiderivative of $Y2(X)$. To do this requires an entirely different approach.

If $f(x)$ is expressed as a power series, than it is a straight forward process to integrate $f(x)$ (symbolically) term by term to determine an antiderivative $F(x)$ of $f(x)$ as a power series. The power series $F(x)$ can then be integrated term by term again to produce its antiderivative $G(x)$ as a power series. This power series process leads to a function $G(x)$ with $G''(x) = f(x)$.

Before we tackle the "rocket" problem below, for which these comments are designed, we try two other problems meant to nudge us gently in this direction.

20. Let $f(x) = x\cos(x^2)$. Evaluate $\int_0^{10} f(x)\,dx$ in the following three ways and compare answers.

    a) Evaluate the integral exactly using analytical means.

    b) Evaluate (approximately) the integral using the $fnInt(\ )$ command.

    c) Write $f(x) = \sum_{j=0}^{\infty} a_j x^j$ as a Taylor series. (Hint: Start with the well known Taylor series for $\cos(x)$ and manipulate it in a straight forward way to produce a Taylor series for $f(x)$. Since a power series can be integrated term by term, we have

    $$\int_0^{10} f(x)\,dx = \sum_{j=0}^{\infty} a_j \int_0^{10} x^j\,dx = \sum_{j=0}^{\infty} a_j \frac{10^{j+1}}{j+1}.$$

    Let $\{s_n\}$ be the sequence of partial sums of the last series. Approximate the value of the integral by evaluating s_n for n sufficiently large. How big does n have to be in order to approximate the known value of the integral (from part (a) with an error less than .0001, and with an error less than 10^{-8}?

21. The function $f(x) = \cos(x^2)$ has a nontrivial antiderivative, which is not available by analytic means. The value of $\int_0^3 f(x)\,dx$ can be computed using the $fnInt(\ )$ command, or by term by term integration of a Taylor series. Compare the two approaches.

22. A rocket, initially at rest, is fired straight up from a mountainous location at an elevation of 12,000 feet above sea level and with an acceleration of $a(t) = t^2 e^{-0.001 t^2}/20$ in ft/sec^2. What is the elevation of the rocket after a 100 second burn? (Use <u>analytic</u> power series methods to determine the elevation function and use your calculator only to compute the appropriate value of the elevation function. To avoid OVERFLOW errors, and to promote greater accuracy, simplify the expression before entering it onto your calculator. Your work should include some evidence concerning the accuracy of your answer. A rigorous error statement is preferred, but other intuitive evidence can be supplied if a rigorous error statement is not included.

Chapter 8

Plane Curves and Polar Coordinates

In this chapter we discuss a new coordinate system and two major new ways that can be used to describe planar curves. These techniques will greatly increase our ability to define a variety of curves. There are important practical consequences, and our work gives us an opportunity to enjoy some truly exotic curves.

8.1 Parametrically Defined Planar Curves

Suppose that f and g are real valued continuous functions of a real variable t defined on an interval $D = [a, b]$. The set, of all points $(x, y) = (f(t), g(t))$ as t ranges over points in D describes a curve C in the plane. The equations $x = f(t)$, $y = g(t)$ are called parametric equations of the curve C, and C is said to be defined parametrically. The variable t is called the parameter. The point $(f(a), g(a))$ is called the initial point, and $(f(b), g(b))$ is called the terminal point. Intuitively, we think of t as time, and $(f(b), g(b))$ as the position at time t of an object moving along the curve. This motion gives the curve an "orientation" from the initial point to the terminal point.

A parametrically defined curve, such as the one described above, can be plotted on our calculator. Press [2ND] [MODE]▼ ▼ ▼ ▶ to select **Par** and press [ENTER]. Press [Y=] and notice the change in the Y= menu, which now has room for six parametrically defined curves: $(X_{1T}, Y_{1T}), (X_{2T}, Y_{2T}), \ldots, (X_{6T}, Y_{6T})$. In addition, pressing the [X,T,θ,n] key enters the letter T (instead of X) onto the active line.

Example 8.1 *Graph the parametrically defined curve* $x = 3\cos(2t)$, $y = \sin(4t)$ *for t in the interval $[0, 2\pi]$.*

We have already put our calculator into its parametric mode by selecting Par on the fourth line of the MODE menu, so there is no need to do this again. Press Y= and enter $X_{1T} = 3*\cos(2*T)$, $Y_{1T} = \sin(4*T)$. In the WINDOW menu, set $T_{min} = 0$ and $T_{max} = 2\pi$. From the formulas involved it is clear that x oscillates between ± 3, and y oscillates between ± 1. We set $X_{min} = -3$, $X_{max} = 3$, and contrary to what might be expected, we also set $Y_{min}=-3$, $Y_{max} = 3$, **because we are more interested here in the actual scaled shape of the curve** than we are in getting the curve to fill up the whole window. For an even more accurate representation, press ZOOM and select Zsquare to see the screen shown.

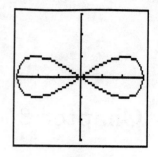

Example 8.2 *Graph the following two parametrically defined curves:*

$$x = 3\cos(2t), \quad y = \sin(4t) \text{ for } 0 \le t \le 2\pi$$

$$x = 4\frac{\cos(t)}{\ln(t)}, \quad y = 4\frac{\sin(t)}{\ln(t)} \text{ for } 3 \le t \le 20.$$

What is notable about this example is that **the parametric range is different for the two curves**. For parametrically defined curves, this is not an uncommon occurrence. To graph both of the curves we must use a common parametric range, and this can be done be reparametrizing one of the curves. We find a linear function $\alpha(t)$ which maps the interval $[0, 2\pi]$ onto the interval $[3, 20]$. All we must do is write the equation of a line through the points $(0, 3)$ and $(2\pi, 20)$, which gives

$$\alpha(t) = 3 + \frac{20-3}{2\pi - 0}(t-0) \Rightarrow \alpha(t) = 3 + \frac{17}{2\pi}t.$$

If t is in the interval $[0, 2\pi]$, then $\alpha(t)$ is in the interval $[3, 20]$. The expressions $X_{1T} = 3*\cos(2*T)$, $Y_{1T} = \sin(4*T)$ are already entered from the last example. A slightly underhanded way can be used to enter $\alpha(t)$. Enter $X_{2T} = 3 + 17/(2*\pi)*T$ and leave Y_{2T} blank. **Notice that the pair X_{2T}, Y_{2T} is automatically in a deactivated state until both X_{2T} and Y_{2T} are given values.** Enter $X_{3T} = 4*\cos(X_{2T})/\ln(X_{2T})$, $Y_{3T} = 4*\sin(X_{2T})/\ln(X_{2T})$. The expression X_{2T} is entered in a way which is similar to the way the familiar $Y1$ is entered. Press VARS ▶ 2 (to select Parametric...), and then 3 to enter X_{2T} at the cursor location on the active line (★86★ x2 can be entered directly from the keyboard. ★86★). All the window variables are already set appropriately from the previous example to produce the screen shown.

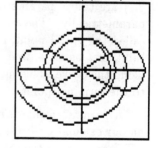

The class of parametrically defined curves is a very broad class of curves, which includes a large variety of new curves that were well beyond our reach without this concept. Surprisingly enough, it also includes all of the familiar curves of the form $y = f(x)$, and significantly, it includes as well all of the curves of the form $x = g(y)$. Here are two examples to demonstrate this versatility.

8.1. PARAMETRICALLY DEFINED PLANAR CURVES

Example 8.3 *Plot, on the same screen, the curve $y = x^2$ ($-3 \leq x \leq 3$) and the parametric curve $x = -1 + 3\cos(t)$ $y = 2 + \sin(t)$ $-\pi/2 \leq t \leq \pi/2$.*

Any curve of the form $y = f(x)$ $(a \leq x \leq b)$ can be defined and graphed parametrically in the form

$$x = t, \; y = f(t) \qquad (a \leq t \leq b).$$

We choose the interval $[-3, 3]$ as our (common) parametric interval. To map this interval onto $[-\pi/2, \pi/2]$, we connect the points $(-3, -\pi/2)$ and $(3, \pi/2)$ with a straight line. This defines

$$\alpha(t) = \frac{\pi}{2} + \frac{-\frac{\pi}{2} - \frac{\pi}{2}}{-3 - 3}(x - 3) = \frac{\pi}{2} + \frac{\pi}{6}(x - 3) = \frac{\pi}{6}x.$$

Enter $X_{1T} = T$, $Y_{1T} = T \wedge 2$, $X_{2T} = -2 + 3\cos(\pi/6 * T)$, $Y_{2T} = 4 + \sin(\pi/6 * T)$. Set $T_{min} = -3$, $T_{max} = 3$, and select ZoomFit to get the screen shown. Notice that under ZoomFit, the values for T_{min} and T_{max} determine the window parameters X_{min}, X_{max}, Y_{min}, Y_{max}. ∎

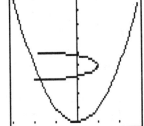

Example 8.4 *Graph the curve $x = y\sin(y)$ ($-10 \leq y \leq 10$) on the usual coordinate system, with the x-axis horizontal and the y-axis vertical.*

Any curve of the form $x = g(y)$ $(c \leq y \leq d)$ can be graphed in the tradition xy-coordinate system by defined it parametrically in the form

$$y = t, \; x = g(t) \qquad (c \leq t \leq d).$$

Enter $X_{1T} = T * \sin(T)$, $Y_{1T} = T$, and set $T_{min} = -10$, $T_{max} = 10$, and select ZoomFit to get the screen shown. ∎

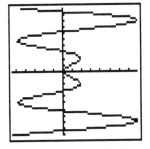

The slope of the line tangent to the graph of the curve C defined parametrically by $x = f(t), y = g(t), (a \leq t \leq b)$ at the point $(x, y) = (f(t), g(t))$ is

$$\frac{dy}{dx} = \frac{\frac{dy}{dt}}{\frac{dx}{dt}} = \frac{g'(t)}{f'(t)}.$$

The arc length differential is

$$ds = \sqrt{dx^2 + dy^2} = \sqrt{\left(\frac{dx}{dt}\right)^2 + \left(\frac{dy}{dt}\right)^2} = \sqrt{(f'(t))^2 + (g'(t))^2}$$

so that the length of a curve C is

$$l(C) = \int_a^b \sqrt{\left(\frac{dx}{dt}\right)^2 + \left(\frac{dy}{dt}\right)^2}\, dt = \int_a^b \sqrt{(f'(t))^2 + (g'(t))^2}\, dt.$$

CHAPTER 8. PLANE CURVES AND POLAR COORDINATES

Parametric curves of the form

$$x = a\cos(mt), y = b\sin(nt), 0 \le t \le 2\pi$$

are known as Lissajous curves. They are well known to provide a wide variety of interesting, complex shapes. In the next example, the subfamily

$$x = a\cos(3t), y = b\sin(2t), 0 \le t \le 2\pi.$$

is considered. Enter $X_{1T} = A * \cos(3 * T)$, $Y_{1T} = B\sin(2 * T)$. Choose a range of values for A and B to see the role these constants play in shaping the curve. To get the screen shown, press [5] [STO▶] [ALPHA] ([A]) [ENTER] and [4] [STO▶] [ALPHA] ([B]) [ENTER], and set the window variables equal to the values $T_{min} = 0$, $T_{max} = 2\pi$, $X_{min} = Y_{min} = -5$, $X_{max} = Y_{max} = 5$.

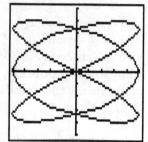

All of the members of this family have a similar shape. A little experimentation quickly confirms, that the value of a controls the amount of horizontal stretch, and the value of b controls the amount of vertical stretch. Each member of the family appears to be symmetric to both the x and y axes, and each of them has several self-intersection points. It seems reasonable to suspect that by choosing a and b appropriately, we can control not only the length of the curve but the angles at some of the self-intersection points. This is the substance of our next example.

Example 8.5 *Consider the family of curves defined parametrically by*

$$x = f(t) = a\cos(3t), y = g(t) = b\sin(2t), (0 \le t \le 2\pi), a, b > 0$$

i) Show that the curves are symmetric with respect to the x-axis and with respect to the y-axis.
ii) Find a member of this family which has an arc length of 20 units, with the additional property that the angles at the self-intersection points off of the coordinate axes are all perpendicular.
iii) Find a point on this curve which is furthest from the origin.

Symmetry with respect to the x-axis is fairly easy to show, because of the well known identities $\sin(-\theta) = -\sin(\theta)$ and $\cos(-\theta) = \cos(\theta)$. If

$$(x, y) = (a\cos(3t), b\sin(2t))$$

denotes a point on the curve, then

$$(x, -y) = (a\cos(3t), -b\sin(2t)) = (a\cos(-3t), b\sin(-2t))$$

is also a point on the curve.

Symmetry with respect to the y-axis is less obvious. We return to the sample graph screen shown above. Press [TRACE], move the trace cursor between points (x, y)

8.1. PARAMETRICALLY DEFINED PLANAR CURVES

and their corresponding symmetry points $(-x, y)$, and observe the corresponding t-values. A little experimentation may lead to the conjecture that the point $(-x, y)$ can be obtained by replacing t by $t + \pi$. This is easy to establish.

$$(a\cos(3(t+\pi)), b\sin(2(t+\pi))) = (a\cos(3t + 3\pi), b\sin(2t + 2\pi))$$
$$= (-a\cos(3t), b\sin(2t)) = (-x, y).$$

There are four self-intersecting points off of the coordinate axes, and because of symmetry we only need to find the one in the first quadrant. We return to the sample graph for guidance, **but we must remember that these points are highly dependent on the values for a and b**, and our sample graph is only one member of the family under consideration. Press [TRACE] and move the trace cursor to the intersection point in the first quadrant, which we denote by (x_0, y_0). (To make more window room, press [2ND] [FORMAT] and select ExprOff.) Make a note of the t-value of the point and then move the trace cursor around the curve until it returns to the same point. From this maneuver, we see that

$$(x_0, y_0) = (f(t_1), g(t_1)) = (f(t_2), g(t_2))$$

for $t_1 \approx .26$ and then again for $t_2 \approx 4.45$.

We could zoom in repeatedly on (x_0, y_0) to find the coordinates of this point with a great deal of accuracy, **but the parameter t does not adjust its values as we zoom in**, so we cannot read off accurate t-values as we zoom in. This is of little consequence, however, since we can certainly deduce accurate values for t_1 and t_2 from accurate values for (x_0, y_0).

We choose, however, a more precise way of getting t_1 and t_2. From the equations

$$(x_0, y_0) = (a\cos(3t_1), b\sin(2t_1)) = (a\cos(3t_2), b\sin(2t_2)).$$

it follows that

$$a\cos(3t_1) = a\cos(3t_2) \text{ and } b\sin(2t_1) = b\sin(2t_2).$$

Since a and b cancel in these equations, it follows that **t_1 and t_2 are independent of a and b**. This important fact needed to be establish anyway, regardless of our approach!

Furthermore, our rough approximations of $t_1 \approx .26$ and $t_2 \approx 4.45$ give us just enough information to actually find their <u>exact values</u>. We know that $\theta_1 = 3t_1 \approx .78$ and $\theta_2 = 3t_2 \approx 13.35$. This means that $0 < \theta_1 < \pi/2$ and $4\pi < \theta_2 < 4\pi + \pi/2$. Since $\cos(\theta_1) = \cos(\theta_2)$, it follows from the figure on the next page that $\theta_2 = \theta_1 + 4\pi$. This gives us

$$t_2 = t_1 + \frac{4\pi}{3}.$$

Now, using the second of the two above equations, we have

$$\sin(2t_1) = \sin(2(t_1 + \frac{4\pi}{3})) = \sin(2t_1 + \frac{8\pi}{3}).$$

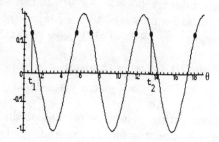

From the *Addition Formula* for the sine function, we have

$$\sin(2t_1) = \sin(2t_1)\cos(\frac{8\pi}{3}) + \sin(\frac{8\pi}{3})\cos(2t_1) = \frac{-1}{2}\sin(2t_1) + \frac{\sqrt{3}}{2}\cos(2t_1).$$

Simplifying gives us

$$\frac{3}{2}\sin(2t_1) = \frac{\sqrt{3}}{2}\cos(2t_1),$$

so that

$$\tan(2t_1) = \frac{1}{\sqrt{3}}.$$

Consequently $2t_1 = \pi/6$ and so t_1 is exactly $t_1 = \pi/12$, and $t_2 = t_1 + 4\pi/3 = 17\pi/12$.

The slope of the tangent line at the point on the curve corresponding to a parametric value of t is

$$m(t) = \frac{g'(t)}{f'(t)} = \frac{2b\cos(2t)}{-3a\sin(3t)} = \frac{b}{a}m_1(t),$$

where

$$m_1(t) = \frac{2\cos(2t)}{-3\sin(3t)}$$

is an expression we can enter on our calculator and evaluate. The two portions of the curve are orthogonal at the self-intersecting point (x_0, y_0) if $m(t_1)m(t_2) = -1$. This gives us the equation

$$-1 = m(t_1)m(t_2) = \frac{b^2}{a^2}m_1(\frac{\pi}{12})m_1(\frac{17\pi}{12}).$$

Enter the formula for $m_1(t)$ as X_{2T} in the Y= menu. We have no intention of using this to create another parametric curve—this is just a convenient way of entering a function. On the home screen, enter $X_{2T}(\pi/12) * X_{2T}(17*\pi/12)$ and press [STO▶] [ALPHA] [M] [ENTER] to store this term as $M = -.6666666667$. The orthogonality statement gives us

$$-1 = \frac{b^2}{a^2}M$$

so that $a = \sqrt{-M}b = .8164965809b$.

This would be a good time to go into the Y= menu and replace A in the formula for X_{1T} by $A = \sqrt{(-M)}B$. Try graphing a few members of our family for various

8.1. PARAMETRICALLY DEFINED PLANAR CURVES

values of B. **In order to check the orthogonality condition, you must choose ZSquare in the ZOOM Menu.**

The relationship between a and b leaves us with only b as an unassigned constant. Arc length is next on our agenda, and the demand to create a curve of arc length 20 will establish a value for b. A formula for arc length was given earlier, and it can be used directly to compute arc length. We still have, however, the unassigned constant B appearing in X_{1T} and Y_{1T} to deal with before we can evaluate the integral. Notice from the formula for arc length that the constant B can certainly be factored out of $f'(t)$ and $g'(t)$, and then factored out of the radical, and then factored out of the integral itself.

A convenient way to accomplish this factoring is to press $\boxed{1}$ $\boxed{\text{STO}\blacktriangleright}$ $\boxed{\text{ALPHA}}$ $(\boxed{\text{B}})$ $\boxed{\text{ENTER}}$ to set $B = 1$. We will calculate the arc length integral and multiply the result by an arbitrary constant b. From the home screen, enter

$$fnInt(\sqrt{(nDeriv(X_{1T}, T, T) \wedge 2 + nDeriv(Y_{1T}, T, T) \wedge 2)}, T, 0, 2*\pi)$$

and press $\boxed{\text{STO}\blacktriangleright}$ $\boxed{\text{ALPHA}}$ $(\boxed{\text{L}})$ $\boxed{\text{ENTER}}$ to get an arc length value of $L = 13.47579067$. Be patient! This will take some time. In order to create a curve of arc length $bL = 20$, we must choose $b = 20/L = 1.484142971$. Store this value to B, and **we have our curve.**

All that is left is to find the point (x, y) on the curve which is furthest from the origin. By plotting this final parametric curve, and moving the trace cursor to the point on the curve which appears to be furthest from the origin, we can see that the maximum distance will occur at a point corresponding to a t-value near $t = 1$.

We can get better graphical evidence. Let $d(t)$ represent the distance from the point $(x, y) = (f(t), g(t))$ on the curve to the origin. **We can graph $d = d(t)$ as an ordinary function of t in a t, d-coordinate system.** With a horizontal t-axis and a vertical d-axis, we might have a better view of the high point on the curve, and the whole problem will certainly be in a more familiar setting. All of our formulas, however, are expressed in terms of T rather than X, and our calculator is currently in parametric mode. Fortunately, as we showed in Example 8.3, it is easy to graph an ordinary function in parametric mode.

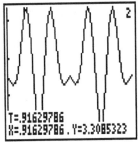

Before we do this, we take advantage of the simple observation that the distance, $\sqrt{x^2 + y^2}$, will be a maximum at the same point that the square of this distance, $x^2 + y^2$ is a maximum. Instead of $d(t)$, we graph $D(t) = (d(t))^2 = x^2 + y^2 = (f(t))^2 + (g(t))^2$. This helps to alleviate a needless complication with square roots. Enter $X_{3T} = T$, $Y_{3T} = (X_{1T}) \wedge 2 + (Y_{1T}) \wedge 2$. Deactivate the pair (X_{1T}, Y_{1T}) and press $\boxed{\text{GRAPH}}$ to see the screen shown. Press $\boxed{\text{TRACE}}$ and move the trace cursor to the high point shown.

We could zoom in on this point, but instead, we solve the equation $D'(t) = 0$ for t near $t = .916$. Press $\boxed{\text{MATH}}$ $\boxed{0}$ to select the SOLVER screen, and enter $eqn : 0 = nDeriv(Y_{3T}T, T)$. Scroll down, and notice that the letters B, T, M involved in this problem all appear in the SOLVER menu. The value for T, which serves as our first guess, is the value obtained from the current trace cursor location. **Remember that the cursor in the SOLVER screen has to be on the line containing the "unknown" letter (in this case T) before we press** [ALPHA] [SOLVE] . The program solves the equation for whatever letter appears on the line where the cursor is located. If you make a mistake and solve for some other letter, the value of that letter will get changed in the process, and some recovery work may be necessary. With the cursor on the line containing T, press [ALPHA] [SOLVE] to get the solution $T = .9424775539004$. It is hard to say how many of these digits are reliable. Recall that the $nDeriv(\ )$ command has an optional argument to initiate greater accuracy, but the accuracy of the output is actually unspecified.

For a more reliable answer we could enter a formula for the exact derivative of $D'(t)$. Given the simplicity of $D'(t)$, one could argue that this is a more appropriate strategy. The details are omitted, but computing and simplifying the equation $D'(t) = 0$, results in

$$6M\cos(3t)\sin(3t) + 4\sin(2t)\cos(2t) = 0,$$

which can be simplified, using trigonometric identities, to

$$3M sin(6t) + 2\sin(4t) = 0,$$

in order to make the equation easier to enter onto the $eqn : 0 =$ line of the SOLVER menu. This more reliable equation leads to a value of $t = T = .94247779608$. Presumably, all or most of these digits are accurate, which would imply that our first approach gave us six digits of accuracy.

To find a point on the curve furthest from the origin, enter X_{1T} and Y_{1T} separately on the home screen to get $(x, y) = (-1.152488062, 1.411503844)$. ■

8.2 Polar Coordinates

Mathematical literature is rich with examples of exotic and beautiful curves which can be drawn in polar coordinates. Some of these appear in the exercise set at the end of the chapter. They should be experienced with delight and enjoyment.

There are two basic ways to plot a polar coordinate equation of the form $r = f(\theta)$. One of them involves using commands we already have available to us. Using the transformation equations

$$x = r\cos(\theta), y = r\sin(\theta)$$

between the polar and rectangular coordinate systems, the polar equation $r = f(\theta)$ can be written in parametric form

$$x = f(\theta)\cos(\theta), y = f(\theta)\sin(\theta),$$

and the curve can then be plotted using the techniques of the previous section. Of course, we would have to use the parametric variable T instead of θ in order to enter the formulas on our calculator.

8.2. POLAR COORDINATES

On the Ti-82, this is the only way to enter a polar coordinate equation. On the Ti-83 and the more advanced models, the calculator can be put directly into polar coordinate mode. Nonetheless, this parametric form is worth keeping in mind, especially if a mix of polar coordinate and parametric curves is considered.

To put our calculator in polar mode, press [MODE] and select **Pol** on the fourth line. Press [Y=] and notice how the menu is set up to enter functions of the form $r = f(\theta)$. The [X,T,θ,n] key now represents the variable θ. We are restricted only to curves of this form. We cannot enter, for example, a curve (admittedly less common) of the form $\theta = f(r)$, but then such restrictions are commonplace. In the familiar function mode, we are restricted to curves of the form $y = f(x)$. (Lest we forget, remember that we can now graph $x = f(y)$ in parametric mode.) The more common polar coordinate curve $r^2 = f(\theta)$ can be graphed as a pair of curves $r = \pm\sqrt{f(\theta)}$.

Discontinuities of the function $r = f(\theta)$ are more difficult to handle in the polar coordinate mode than discontinuities of $y = f(x)$ were in the more familiar function mode. Graphing a discontinuous function $y = f(x)$ under ZoomFit sometimes produced a large scale on the y-axis. This frequently required special handling, but the adjustments were usually straight forward. In polar coordinates, graphing discontinuous functions under ZoomFit can produce "large window" problems, which are not always easy to interpret.

If the equation $r = f(\theta)$ is written parametrically, then there is no need to depend on any special forms for the slope of the tangent line, or for the arc length differential. **This is a distinct advantage of writing polar coordinate curves in parametric form.** We simply use

$$\frac{dy}{dx} = \frac{\frac{dy}{d\theta}}{\frac{dx}{d\theta}}, \quad ds = \sqrt{\left(\frac{dx}{d\theta}\right)^2 + \left(\frac{dy}{d\theta}\right)^2}.$$

However, if desired, these forms readily lead to the formulas

$$\frac{dy}{dx} = \frac{f'(\theta)\sin(\theta) + f(\theta)\cos(\theta)}{f'(\theta)\cos(\theta) - f(\theta)\sin(\theta)},$$

$$ds = \sqrt{r^2 + \left(\frac{dr}{d\theta}\right)^2}\, d\theta = \sqrt{(f(\theta))^2 + (f'(\theta))^2}\, d\theta.$$

The arc length of the curve $r = f(\theta)$ on $(\alpha \leq \theta \leq \beta)$ is

$$Length = \int_\alpha^\beta \sqrt{(f(\theta))^2 + (f'(\theta))^2}\, d\theta.$$

If R is a region bounded by $r = f(\theta)$, $\theta = \alpha$, and $\theta = \beta$ with $\alpha < \beta$, then the area of R is

$$Area = \int_\alpha^\beta \frac{1}{2}(f(\theta))^2\, d\theta.$$

The details of all of these formulas, and others, will be found in any standard calculus text.

Example 8.6 *Find the area of the region which lies inside the four-leaved rose* $r = 7\cos(2\theta + 1)$ *and above the horizontal line* $y = 3$.

Using the polar transformation equation $y = r\sin(\theta)$, the equation $y = 3$ can easily be expressed in the form $r\sin(\theta) = 3$ and then in the form $r = 3/\sin(\theta)$. Unfortunately, discontinuities are involved in this form, which will cause precisely the kind of "large window" problem we referred to above, if we try to plot this pair of equations under ZoomFit. One way around this problem is to graph both curves parametrically. This is, after all, a more natural way to define the curve $y = 3$. On the other hand, polar coordinates can still be used, but we must control window size more aggressively. This is the approach we use.

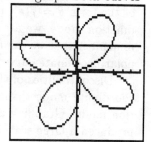

With our calculator in polar coordinate mode, we enter $r1 = 3/\sin(\theta)$, $r2 = 7*\cos(2*\theta+1)$, and set $T_{min} = 0$, $T_{max} = 2\pi$. Considering the curves involved, the values of $X_{min} = -7$, $X_{max} = 7$, $Y_{min} = -7$, $Y_{max} = 7$, are natural choices. Avoiding the use of ZoomFit produces the screen shown, which clearly avoids any window problem created by the discontinuity.

The graph of the region shows that there are four points of intersection. Our calculator is in polar graphing mode, **but it is not yet in polar coordinate mode.** Before we can find the polar coordinates of these points of intersection, we must press [2ND] [FORMAT] and select **PolarGC**. (★86★ The "Pol" mode for polar graphing, and the "PolarC" mode for displaying polar coordinates, are both in the MODE menu. ★86★) Once this is selected, press [TRACE] and move the trace cursor to the various intersection points on the curve. The results are surprising. This is so typical! The polar coordinates of a point on a curve have a wide range of very different values, some of which may not satisfy the equation. The trace cursor gives us approximate values of $(r, \theta) \approx (3.21, 2.09)$, $(6.83, 2.75)$, $(-4.71, 3.80)$, $(-3.70, 4.75)$. We will always get coordinates which satisfy the equation from the trace cursor, but judging from the last two pairs, they may be different from what we expected. We will use the θ-values in the SOLVER menu to get more accurate values, and so we store these values as $A = 2.09$, $B = 2.75$, $C = 3.80$, $D = 4.75$. Press [MATH] [0] to select the SOLVER menu, and enter $eqn : 0 = r1 - r2$. (The expression $r1$, for example, is entered in a way which is similar to the way the familiar $Y1$ is entered. Press [VARS] ▶ [3] (to select Polar...), and then [1] to enter $r1$ at the cursor location on the active line.) (★86★ Expressions like $r1$ can always be entered directly from the keyboard. ★86★) After the equation is entered, press ▼, enter $\theta = A$, and press [ALPHA] [SOLVE]. When a value is returned press [2ND] [QUIT] [X,T,θ,n] [STO▶] [ENTER] to reset the value of A more accurately as $A = 2.119291465$. Return to the SOLVE screen and repeat these steps, letting $\theta = B$, C, and then D, to get the more accurate values of $B = 2.695870991$, $C = 3.809252544$, $D = 4.775853085$.

In order to use the polar coordinate area formula effectively, it is important to have a good geometric interpretation of the various parts of the area formula $\int_\alpha^\beta \frac{1}{2} r^2 \, d\theta$. In the derivation of the formula, we identify $\frac{1}{2} r^2 \, d\theta$ with the area of a

8.2. POLAR COORDINATES

"thin" circular sector located at angle θ. The integral symbol $\int_\alpha^\beta$ represents the sum of all of these terms as θ goes from α to β. In the figure shown on the next page, a typical sector for the curve $r = 7\cos(2\theta + 1)$ is shown for θ in the interval $A \leq \theta \leq B$.

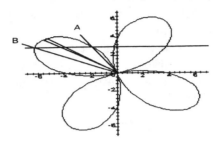

This interpretation, and the picture it creates, leads immediately to a formula for the area of the portion of the region in the second quadrant. From the home screen, we enter $fnInt((r2 \wedge 2 - r1 \wedge 2)/2, \theta, A, B)$ and press [STO▶] [ALPHA] ([u]) to get $U = 3.714198476$. (⋆86⋆ You are encouraged to use more natural names. ⋆86⋆) The area of the second region is just as easy to compute, though the parameters involved may initially cause concern. The negative r-values are acceptable, because the squaring process in the integration formula will return positive values. The angles $C \leq \theta \leq D$ are almost entirely in the third quadrant. This may be disconcerting, but such angles, along with negative r-values are quite acceptable. We enter $fnInt((r2 \wedge 2 - r1 \wedge 2)/2, \theta, C, D)$ and press [STO▶] [ALPHA] ([v]) to get $V = 11.28274411$. (Incidentally, **the easiest way to enter this second integral** is to press [2ND] [[ENTRY]] to bring the first area statement back to the active line, where it can be easily edited to produce the second formula.) To find the area of the region which lies inside of $r = 7\cos(2\theta + 1)$ and above the horizontal line $y = 3$, we press [ALPHA] ([u]) [+] [ALPHA] ([v]) [ENTER] to get $U + V = 14.99694259$. ∎

Example 8.7 *Graph the polar coordinate curve*

$$r = 3\sin(7\theta/3).$$

Find the angle between the intersecting branches of the curve at one of the self-intersecting points (other than the origin).

The MODE and FORMAT menu settings from the last example should be maintained. Let $r = f(\theta) = 3\sin(7\theta/3)$. **Graphing this curve on the interval $0 \leq \theta \leq 2\pi$ will not produce a complete graph.** How should we choose α so that graphing $r = f(\theta)$ on the interval $[0, \alpha]$ will produce a complete graph? We would probably choose α to be a multiple of 2π in order to make a complete circuit around the origin. Also α should probably be a multiple of the period of $f(\theta)$ so the the r-values repeat themselves. The period p satisfies $\frac{7p}{3} = 2\pi$, which gives $p = \frac{6\pi}{7}$. From this analysis, we choose $\alpha = 6\pi$.

We let $r1 = 3*\sin(7*\theta/3)$, and set $\theta_{min} = 0$, $\theta_{max} = 6*\pi$. Set the remaining WINDOW parameters equal to ±3, and select ZSquare to produce the screen shown. This global picture doesn't show the intersecting loops very well, but it is a better picture for studying coordinate values. Press [TRACE] and observe the coordinates as the trace cursor moves around the curve. Much to our surprise we discover that the curve is completely plotted on the interval $[0, 3\pi]$—only half of the plotting interval we selected. The reason for this can be seen by observing the coordinates as the trace cursor moves beyond 3π. Evidently,

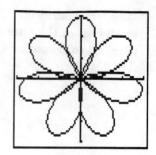

$$r = f(\theta + 3\pi) = -f(\theta).$$

This conjecture is easily proved, and it explains why the shorter plotting interval is sufficient.

There are only two ways in which the curve could self-intersect. One way would be at a value θ_0 for which

$$r_0 = f(\theta_0) = f(\theta_0 + 2\pi),$$

and the other at a value θ_0 for which

$$r_0 = f(\theta_0) = -f(\theta_0 + \pi).$$

A solution to either one of these equations would produce a self-intersecting point. By moving the trace cursor around the curve, it is readily seen that both types of intersections occur. The self-intersection point at $(r_0, \theta_0) \approx (1.5, .25)$ satisfies the first of these equations, and this equation can be used on our calculator to find more accurate values for (r_0, θ_0). Press [MATH] [0] to select the SOLVER menu, and enter the equation $eqn: 0 = r1 - r1(\theta + 2*\pi)$, press ▼ and use $\theta = .25$. Press [ALPHA] ([SOLVE]) to get $\theta_0 = .22439947525637$. As an alternate strategy, this equation could have been solved exactly for θ_0 using trigonometric identities, to get $\theta_0 = \pi/14$. The details are not included, but notice that $\pi/14 = .2243994753$.

To find the angle between the intersecting branches of the curve we will use the trigonometric identity

$$\tan(\alpha - \beta) = \frac{\tan(\alpha) - \tan(\beta)}{1 + \tan(\alpha)\tan(\beta)}.$$

The slope of any line is the tangent of the angle of inclination of the line. A slope formula in terms of polar coordinates was given in the introduction of this section.

We intend to enter this slope formula onto our calculator. First we compute $f'(\theta) = 7\cos(\frac{7\theta}{3})$, and enter this as $r2$. It may seem to be inappropriate to use $r3$ for our slope formula, but it is a means of entering a function of θ. Enter

$$r3 = (r2*\sin(\theta) + r1*\cos(\theta))/(r2*\cos(\theta) - r1*\sin(\theta)).$$

Let α and β be the angles of inclination at the self-intersection point with θ-values of θ_0 and $\theta_0 + 2\pi$, respectively. (As long as we haven't changed the value of

8.3. EXERCISE SET

θ, it continues to be $\theta = \theta_0$.) Then, using the above trigonometric identity, we find the value of $\tan(\alpha - \beta)$ by entering

$$(r3 - r3(\theta + 2*\pi))/(1 + r3*r3(\theta + 2*\pi)).$$

to get $\tan(\alpha - \beta) = .527145898$. Press [2ND] [TAN^{-1}] [2ND] [ANS] [ENTER] to get an angle between the intersecting branches of the curve at the self-intersecting point of $\alpha - \beta = \tan^{-1}(.527145898) = .4851277482$ in radian measure. ∎

Example 8.8 *Graph the ellipse defined by the polar coordinate equation*

$$r = f(\theta) = \frac{30}{13 - 12\sin(\theta + \frac{\pi}{5})},$$

and find its arc length.

The graph is straight forward. To compute arc length, we will need the derivative of $r = f(\theta)$. We could use the $nDeriv(\ )$ command, but our result will be more reliable, if we compute an exact formula for the derivative. There is a way to express $f(\theta)$, which will make all of this easy to enter onto our calculator. Let $g(\theta) = 13 - 12\sin(\theta + \frac{\pi}{5})$, so that $f(\theta) = 30 * (g(\theta))^{-1}$. In this form we can write $f'(\theta) = -30(g(\theta))^{-2}g'(\theta)$.

Enter $r1 = g(\theta) = 13 - 12*\sin(\theta + \pi/5)$, $r2 = 30/r1$, $r3 = g'(\theta) = -12*\cos(\theta + \pi/5)$, and finally $r4 = -30 * r1 \wedge (-2) * r3$. Deactivate all of the expressions except $r2$ which is the ellipse itself. Set $\theta_{min} = 0$, $\theta_{max} = 2*\pi$, to get the screen shown under ZoomFit.

Our preparations make the computation of arc length an easy task. A formula for arc length was given in the introduction. From the home screen, enter

$$fnInt(\sqrt{(r2 \wedge 2 + r4 \wedge 2)}, \theta, 0, 2*\pi)$$

to get an arc length of 71.25302208. ∎

8.3 Exercise Set

1. Parametric equations of the form

$$x = x_0 + a\cos(mt), y = y_0 + b\sin(mt), 0 \le t \le 2\pi$$

are effective ways of representing a circle or an ellipse having a horizontal or vertical major axes. Ultimately these parametric equations should become so familiar to a student of mathematics that calculators are not needed to plot them or grasp their significance. Use pencil and paper techniques, and a calculator to help gain some insight into these forms. Determine the role played by x_0, y_0, by a, b, and by m. Think of the parametric curve as describing motion of an object along the curve in terms of time t.

2. A straight line in the plane is represented parametrically by

$$x = x_0 + pt, y = y_0 + qt, -\infty < t < \infty,$$

but parametric equations of straight lines do not have to be linear. Use your calculator to plot

$$x = 3 + 2f(t), y = 5 - 3f(t), t_0 < t < t_1$$

for the following sample of functions f and intervals $[t_0, t_1]$.
  a) $f(t) = t$ on $(-\infty, \infty)$ b) $f(t) = t^3$ on $(-\infty, \infty)$
  c) $f(t) = t^2$ on $(-\infty, \infty)$ d) $f(t) = \tan(t)$ on $(-\frac{\pi}{2}, \frac{\pi}{2})$
What role does the function f play in these representations? Experiment further, if necessary, to answer this question. Think of the parametric curve as describing motion of an object along the curve in terms of time t.

3. Plot the Hypotrochoid

$$x = \cos(t) + 5\cos(3t), y = 6\cos(t) - 5\sin(3t), 0 \le t \le 2\pi.$$

4. A circle of radius 1, with a fixed point P marked on the circle, lies inside a circle of radius 10 centered at the origin, and is allowed to roll along the circumference of the larger circle. Initially, the point P is at the point (10,0) and the small circle rolls in a counterclockwise direction along the inside of the larger circle. The point P traces out a curve defined parametrically by

$$x = 9\cos(t) + \cos(9t), y = 9\sin(t) - \sin(9t), 0 \le t \le 2\pi.$$

Use your calculator to plot this curve.

5. Parametric curves of the form

$$x = a\cos(mt), y = b\sin(nt), 0 \le t \le 2\pi,$$

are known as Lissajous curves. Use your calculator to graph a sample of them and to determine the role played by each of the constants, a, b, m, n.
a) Fix $n = 2, m = 3$, and plot over a range of a, and b values.
b) Plot with $a = 2, b = 5, n = 1$ fixed, and $m = 2, 3, 6$.
c) Plot with $a = 2, b = 5, n = 2$ fixed, and $m = 3, 5, 7$.
d) Plot with $a = 2, b = 5$ fixed, and $n = 2s, m = 3s$ for $s = 2, 3, 4$. Notice that a common factor, j, between $n = n_1 j$ and $m = m_1 j$ has no effect on the graph. Why? If we happen to know the basic shapes of these curves for all integers m, and n, is there anything gained by considering rational numbers m, and n, which are not integers?

6. a) Plot the rose petal curve defined by the polar coordinate equation $r = 3\cos(2\theta)$.
b) Plot the polar coordinate curves $r = 3\cos(2\theta + s)$ for $s = 0.3, 0.6, 0.9, 1.2$. Describe, in specific terms, the role played by s.
c) Prove your claim in part (b) analytically.

8.3. EXERCISE SET

7. Plot the following conic sections in polar coordinates and find the vertices and the second focus in polar and rectangular coordinates. Graphing problems caused by infinite discontinuities should be dealt with effectively.

   a) $r = \frac{4}{3+2\cos(\theta - \frac{3\pi}{4})}$ b) $r = \frac{8}{3-3\cos(\theta + \frac{\pi}{6})}$ c) $r = \frac{21}{5+7\cos(\theta+1)}$

8. Consider the family of curves defined by $r = \sin(m\theta)$. The value of m has a dramatic effect on the graph.

   a) When m is an positive integer the curves are called "rose petal" curves. Plot these curves for $n = 2, 3, 4, 5, 6, 7$. Make a conjecture about value of the integer m and the shape of the curve. Explain, analytically, why the patterns are different for m even or odd.

   b) Plot the curves for $m = 5/3$, and then for $m = 7/2, 7/3, 7/4, 7/6, 7/11$. Any conjectures?

   c) A larger denominator seems to complicate the curve. What would be the result be if m were irrational? Try to answer this question first, then plot the curve for $m = \sqrt{2}$. Use a large θ-range. Why? Then sit back, and relax. This will take some time.

   d) What about curves of the form $r = \cos(m\theta)$? Do we have to go through all this work again? Think about Problem 9 and an elementary trigonometric identity.

9. The *folium of Descartes* is defined by the equation $x^3 + y^3 = 3xy$. Graph this curve by converting it to polar coordinates. Use the connected dot style (the standard style) of graphing. Recall that the back slash just to the left of $r1$ in the Y= menu controls the graphing mode. (★86★ Look for the Style subsubmenu of the $r(\theta) =$ submenu of the GRAPH menu. ★86★) Carefully comment on the result, and explain why the calculator behaves as it does. Fix the problem, and graph the curve again. It will help to set θ_{step} equal to a smaller number, although this will increase the amount of time it takes to produce the curve.

10. Plot the beautiful **Fay Butterfly**

$$r = e^{\cos(\theta)} - 2\cos(4\theta) + \sin^5(\frac{\theta}{12}).$$

Notice that a θ-interval larger than 2π must be used. How big must the plot range be in order to display the whole curve? Choose a range of minimal length which will produce the entire curve, and carefully argue why the range you choose is sufficient. Plot the curve again using the unnecessarily large range of $[0, 1000]$, and notice how poorly the curve is plotted. Explain why your calculator behaves in this way. This curve appears in "The Butterfly Curve" by Temple H. Fay, Amer.Math.Monthly **96** No.5, May 1989, pp.442-443.

This curve will take some time for your calculator to draw, but be patient. The result is worth the wait—enjoy! Incidentally, the intricate lacy filigree can be fully appreciated only by viewing this curve through some plotting program on a desk top computer.

11. Equations in x and y are often difficult to graph, and equally difficult to manage in other ways as well. Frequently, we try to find a parametric representation for the curve defined by the equation. This not only allows us to graph the curve on our

calculator, but it gives us many other advantages as well. We have seen in this chapter how parametrically defined curves can be analyzed and manipulated.

The techniques we use to find a parametric representation for a curve defined by an equation in x and y are limited only by our imagination. Frequently it is some geometric insight that leads to a formula, but once we find a formula, the geometry can be discarded. What we enter on our calculator, after all, is simply a formula without any geometric meaning attached to the parameter. Graph the curves defined by the following equations by finding a parametric representation for each of them. Hints are given, but think of the hints only as examples. Perhaps you can find some other way to parametrize the curve.

a) Hint: Complete the squares and try a trigonometric substitution. Use the identity $\cos^2(t) + \sin^2(t) = 1$.
$$5x^2 + 9y^2 - 3x + 2y = 87$$

b) Hint: This is somewhat similar to part a). Use the identity $\cosh^2(t) - \sinh^2(t) = 1$ instead.
$$8x^2 - y^2 + 3x + 5y = 13$$

c) Hint: The lines $y = mx$ for $-\infty < m < \infty$ sweep out all the points in the plane except the points $y \neq 0$ on the y-axis. Think of m as a parameter. Find the point of intersection (x, y) between this curve and the line $y = mx$. Express x and y in terms of m.
$$x^4 + y^4 = 14x^3 + 5y^3$$

d) Hint: The parabolas $x = my^2$ for $-\infty < m < \infty$ sweep out all the points in the plane except the points $x \neq 0$ on the x-axis. Think of m as a parameter. Find the point of intersection (x, y) between this curve and $x = my^2$. Express x and y in terms of m. (Given the solution to part (c), can you see how we are naturally lead to this approach?)
$$x^2 + y^4 = 37xy$$

12. The class of parametrically defined curves includes a large supply of exotic and beautiful curves. Most calculus texts have an abundance of examples. The following curve, which is particularly beautiful, appears in "Wheels on Wheels on Wheels—Surprising Symmetry" by Frank A. Farris, Mathematics Magazine, **69** No.3, June 1996, pp. 185-189. Plot this curve, defined below, and find its arc length. **The value of T_{step} in the WINDOW menu must be changed to a small number in order to plot this curve effectively.** With a smaller value of T_{step}, the curve will take much more time to graph. A value of $T_{step} = .01$ is suggested, and the curve should be plotted under ZSquare.

$$x = \cos(t) + \frac{1}{2}\cos(7t) + \frac{1}{3}\sin(17t)$$

$$y = \sin(t) + \frac{1}{2}\sin(7t) + \frac{1}{3}\cos(17t)$$

13. Plot the curve $x = t^2 - t$; $y = 2t^2 - t^3 + 7$ in a window which shows its self-intersection point. Find the coordinates of the point of intersection expressed

as decimal numbers. Find the angle (acute angle) between the two intersecting subarcs at the intersection point. Express your answer in degree measure. Find the length of the arc (loop) beginning and ending at this intersection point.

14. Consider the torus generated by rotating, about the x-axis, the circle of radius 2 centered at $(x_0, y_0) = (0, 6)$. Find the surface area of the torus. Use a parametric form for the equation of the circle.

15. Find the area of the region inside the graph of $r = -4\sin(t)$, and outside the graph of $r = 6\sin(3t)$.

16. Find the area of the region outside of the graph of the cardioid $r = 2 + 2\cos(\theta)$ and inside the polar coordinate curve $r^2 = 25\cos(\theta)$. Plotting the second curve will require some special handling. The restricted domain of the second curve will also cause some complications. A full plot of the region involved should be included.

Project:Half of a Rose

Find a horizontal line $y = a$ above the x-axis, such that the area of the region above this line and inside the three-leaved rose $r = 5\sin(3\theta)$ is half the area of the entire rose.

This problem, which resembles Example 8.6, seems simple enough. The area of the entire rose is easy to compute. Then, all one has to do is find the points of intersection of the line and the rose, and the rest is fairly straightforward. An exact solution to the equation involved in this point of intersection, however, is too complicated to be useful, and so it is natural to look for a solution with the help of the SOLVE menu. Unfortunately, the value of a that controls the horizontal line is unknown, and in order to solve an equation, every name in the equation must evaluate to a numeric except for the unknown that the equation is being solved for.

The easiest way to proceed is by guess work. Guess at an answer, compute the area involved, use this to adjust your guess. This can be repeated again and again, each step leading to a better answer.

A better way is to write a program which mimics your guess work. Perhaps you can think of an approach that works better than the one discussed below. Notice, however, how "natural" the idea is, how methods of this sort can be created without any background in numerical approximation techniques.

Start with a value for a in $y = a$, call it H (for HI), which is too big, and another value for a, call it L (for LOW), which is too small. Clearly $a = H = 4$, and $a = L = 0$ will work. Let M (for MID) be the average of L and H. One can then check the area corresponding to $a = M$ to see if it is larger or smaller than half the area of the rose. Use this to reassign either L or H to be the value of M. This would complete one pass through the program loop, and the steps are then repeated as often as necessary until $a = M$ is close enough.

Write a program which solves this problem. To avoid integrating each pass through the program loop, **integrate just once by hand, and generate a formula** that can be used instead to evaluate area in each pass through the program loop.

Your calculator will have to be told to stop the program when the answer is close enough. To do this in a program, the word **"while"** is used in the following way:

:Initial program steps which, among other
:things, determine the error term E.
:*while* $E > 10 \wedge (-8)$ This command is in the CTL
                                     submenu of the PRGM menu.

:Various commands to do
:under this condition
:*end*   This command, in the CTL submenu of the
                                     PRGM menu, ends the "while" group of terms.

The expression E is a measure of how close the approximate answer is to the correct answer. There are several ways to measure this, and the choice is left to you. As soon as *error*$\leq 10^{-8}$ the program should be written to stop.

Project: Working on the Railroads

A traffic control engineer working for a railroad corporation watches the progress of two trains on a computer console. The trains are traveling on different tracks which intersect at a point ahead of both trains. The coal train is 1.57 miles long and the position of its head (its front most point) at time t is

$$x = 17t - 4\cos(2t); y = 2t + 3\sin(5t).$$

The freight train is 1.14 miles long and the position of its head at time t is

$$x = 30 + 11t, y = 117 + 18t - 7t^2.$$

In both cases t represents time, in hours, and x and y are in miles. The engineer makes a quick calculation and decides that the trains will safely pass through the intersection point. Unfortunately, this is not a good decision.

a) Show conclusively that the trains collide, and determine which train crashes into the other.

When the mess is cleaned up, management decides that it better send all of its traffic control engineers to school to learn some calculus. When the course is completed each engineer will be given a test. Failure will mean immediate job dismissal!

Let us tune back in to this drama, then, after some time has passed, and suppose that the moment has arrived. As one of the traffic control engineers, here is the test that you must now take.

b) Given the exact circumstances surrounding the accident, what is the minimal amount of extra time that the coal train must have, in order for it to pass the intersection point (crash free) just ahead of the freight train? What is the minimal amount of extra time that the freight train must have, in order for it to pass the intersection point (crash free) just ahead of the coal train?

8.3. EXERCISE SET

If one of the trains had exactly this amount of extra time, then the trains would safely pass through the intersection point of the two railroad tracks, but the trains would come so close to each other, that the experience would clearly scare everyone involved. Consequently, you are now asked to add a margin of safety to your calculations, so that the trains will "safely" pass each other. Think of two distinctly different ways of adding a safety margin, describe them carefully, and recalculate answers with a built-in safety margin.

Part (b) (the test) is not easy. Each answer is the solution to an arc length equation.

The easiest way to proceed is by guess work. Using the results from part a), you know a t-value which is too small (the trains crash). By guess-work, find a t-value which is too big (the trains do not crash). It follows that the correct answer is some t-value in between. Guess again, and decide whether your guess is too big or too small. Now what interval is the correct answer in? Guess again, and again. With each guess, the correct answer is narrowed down to a smaller interval.

A better approach would be to write a program which mimics all of this guess work. Start with a t-value, call it L (for LOW) which is too small, and a t-value, call it H (for HI) which is too big. Let M (for MID) be the average of L and H. Decide whether $t = M$ is too big or too small, and use this to reassign either L or H to be the value of M. This would complete one pass through the program loop, and the steps are then repeated as often as necessary until M is close enough.

Your calculator will have to be told to stop the program when the answer is close enough. To do this in a program, the word **"while"** is used in the following way:

:Initial program steps which, among other
:things, determine the error term E.
:while $E > 10 \wedge (-8)$ This command is in the CTL
                                         submenu of the PRGM menu.

:Various commands to do
:under this condition
:end   This command, in the CTL submenu of the
                                         PRGM menu, ends the "while" group of terms.

The expression E is a measure of how close the approximate answer is to the correct answer. There are several ways to measure this, and the choice is left to you. As soon as $error \leq 10^{-8}$ the program stops.

Chapter 9

Vectors and Analytic Geometry

A vector is a directed line segment, or more precisely, a class of all directed line segments having the same length and direction. Soon after addition and scalar multiplication are defined geometrically, we arrive at the familiar form $\vec{v} = a\vec{i} + b\vec{j}$ for a vector in two-dimensional space, and $\vec{w} = a\vec{i} + b\vec{j} + c\vec{k}$ for a vector in three-dimensional space, where $\vec{i}, \vec{j}, \vec{k}$ are the standard unit vectors along the positive coordinate axes. These forms can be reduced further to the ordered pair $\vec{v} = (a, b)$ and the ordered triplet $\vec{w} = (a, b, c)$.

We take up the subject with this symbolic vector form. A vector is simply an ordered pair or ordered triplet of real numbers. A point in the plane or in space is, of course, denoted in the same way. Indeed, a point and a vector are two different models for the same abstract mathematical structure. This double meaning may seem confusing initially, but it turns out to be a blessing. In practice, we switch our point of view from a point to a vector and back again to suit our purpose. Used in this way, **vectors can be a powerful tool for solving a wide range of mathematical problems**.

Vectors do not have an official home inside of our calculator, but it is very easy for us to make a place for them. (⋆86⋆ Just the opposite is true on this device. Press [2ND] [VECTR] to access a full menu of commands. ⋆86⋆) A "list" on our calculator is an ordered collection of numbers, and so it is a perfect mechanism for representing vectors.

The vector $\vec{v} = a\vec{i} + b\vec{j}$ can be entered onto our calculator by entering $\{5, -7\}$ and storing the result to V. Press [2ND] [{] [5] [,] [(−)] [7] [2ND] [}] [STO▶] [ALPHA] [v] [ENTER]. (⋆86⋆ The curly bracket symbols are in the LIST menu. A "vector" (as opposed to a "list") is entered in a different way on this device. However, it will frequently be necessary to represent a vector as a list. Our representation here of a vector as a list should not be ignored.⋆86⋆) Vectors of the form $\vec{w} = a\vec{i} + b\vec{j} + c\vec{k}$ are entered in a similar way. The vector $\vec{v} = 5\vec{i} - 7\vec{j}$, as we entered it above, can be used with other vectors in 2-space, but $\vec{v}$ can also be regarded as a vector in 3-space—after all, the plane is a subset of 3-space. In this case, to make $\vec{v}$ compatible with other vectors in 3-space (rather than 2-space), it must be entered in the form $\{5, -7, 0\}$. Adding the extra 0 would be a critical matter in this situation, if a calculator is to be used.

We worked with lists in earlier chapters, and in order to understand the role played by our calculator in creating and using lists, a review of Sec 4.1 and Sec 7.1 is suggested. A moment ago, we stored the vector $\vec{v} = 5\vec{i} - 7\vec{j}$ in memory to V. In order to use the vector $\vec{v}$ (which is actually a list) for further work on our calculator, recall that **we must put the prefix L in front of V and enter the vector in the form $_LV$.** (⋆86⋆ The prefix L is unnecessary—it doesn't exist—on this calculator. If a list is stored to memory V, then the list V can be used on the calculator by simply entering the letter V. ⋆86⋆) The prefix L can be entered by pressing [2ND] [LIST] ▶ (to get the OPS submenu) and then scrolling down to select the last item or pressing [ALPHA] ([B]). **An alternate way, however, is usually more convenient.** Press [2ND] [LIST] and scroll down through the NAMES submenu. The list V we created above will appear in this submenu. Select V and press [ENTER] to enter $_LV$ onto the active line. **Notice that the prefix L automatically appears in front of V.**

An even more convenient way to access a vector, after it has been created, is to store it to one of the keyboard registers $L_1, L_2, \ldots, L_6$. These locations on the keyboard —in gold print—are reserved strictly for lists. They are particularly easy to use not only because they are located directly on the keyboard, but also because the prefix L does not have to be attached to them. The disadvantages to using them are obvious. There are only six of them, and by using them, we are passing up the opportunity to use names which are more motivating. Names like $\vec{v}$ for velocity and $\vec{a}$ for acceleration can often make a problem seem simpler.

Finally, the components of a vector $_LV$ (or any list for that matter), are extracted in a familiar way. To determine the second component of $_LV$, for example, we enter $_LV(2)$.

⋆⋆⋆⋆86⋆⋆⋆⋆

Press [2ND] [VECTR] to access the VECTOR menu. A vector is an ordered list of numbers enclosed in square brackets ([...]). A "list" is an ordered list of numbers enclosed in curly brackets ({...}). **The two data types are very different, but both can and will be used to represent vectors.** Both the VECTOR menu and the LIST menu have commands for converting one data type into the other. These two conversion commands will be useful. Most of Chapter 12 on vectors in the Ti-86 guidebooK should be read.

Vectors can be entered from the home screen or from the VECTOR EDIT submenu. To enter the vector $\vec{v} = 5\vec{i} - 7\vec{j}$ from the home screen, press [2ND] [[] [5] [,] [(−)] [7] [2ND] []] [STO▶] [ALPHA] ([v]) [ENTER] to store the vector as V. Small case letters can be used as well, and the letters V and v can be used to store two different vectors. To enter $\vec{v}$ from the VECTOR menu, press [2ND] [VECTR] [F2] (to select the Edit submenu). The calculator will be in locked alpha mode so that a name, in our case V, can be entered conveniently. Press [ENTER] and the cursor moves to a location where the dimension of the vector can be entered. The vector we are entering has a dimension of 2, so press [2] [ENTER] and the cursor moves to a location on the screen where the components of the vector can be entered in a obvious way. Components automatically have 0-values until the desired values are entered.

Once they are entered into memory, vectors and lists are easy to use in other computations. Neither data type requires any special prefix on this machine. Vectors and lists can be entered onto any screen by entering the appropriate name from the keyboard, or by pasting the name through the NAMES submenu of the LIST or VECTOR menu.

★★★★86★★★★

9.1 Addition and Scalar Multiplication

One of the most useful properties of lists on our calculator is that they are "listable." Generally speaking, this means that if you perform an operation on a list, the result is the list obtained by performing the operation on the elements of the list. **This property allows us to perform algebraic operations with vectors in a very natural way.**

Example 9.1 *Compute the vectors $\vec{a} = \vec{u} + \vec{v}$, $\vec{b} = \vec{v} - \vec{w}$, $\vec{c} = 7\vec{v}$, and $\vec{d} = 3\vec{u} - 4\vec{v} + 6\vec{w}$, if $\vec{u}, \vec{v}, \vec{w}$ are the vectors defined by*

$$\vec{u} = 2\vec{i} - 3\vec{j} + \vec{k}, \ \vec{v} = 6\vec{i} + 7\vec{k}, \ \vec{w} = \vec{i} + 5\vec{j} - 8\vec{k}.$$

Press [2ND] [[ ] [2] [,] [(-)] [3] [,] [1] [2ND] [ ]] [STO▶] [2ND] [L1] [ENTER] to store the vector $\vec{u} = \{2, -3, 1\}$ to L_1. In the same way, enter $\vec{v} = \{6, 0, 7\} = L_2$, and $\vec{w} = \{1, 5, -8\} = L_3$.

It is now an easy matter to compute the desired vectors. To compute $\vec{a}$ simply enter $L_1 + L_2$ by pressing [2ND] [L1] [+] [2ND] [L2] [ENTER] to get

$$\vec{a} = \{8, -3, 8\} = 8\vec{i} - 3\vec{j} + 8\vec{k}.$$

In a similar way we enter

$$\vec{b} = L_2 - L_3 = \{5, -5, 15\} = 5\vec{i} - 5\vec{j} + 15\vec{k}, \ \vec{c} = 7 * L_2 = \{42, 0, 49\} = 42\vec{i} + 49\vec{k},$$

and

$$\vec{d} = 2 * L_1 - 4 * L_2 + 6 * L_3 = \{-14, 24, -74\} = -14\vec{i} + 24\vec{j} - 74\vec{k}.$$

(★86★ Use the [+] and [-] keys in the same way to perform vector addition and scalar multiplication regardless of whether vectors are represented on the machine as "lists" or as "vectors". ★86★) ■

The listable property of lists also allows us to perform silly computations like $\vec{u} + 5 = L_1 + 5 = \{7, 2, 6\}$. We get a value, but it obviously makes no sense as a vector valued concept. The lack of an error message should not be taken as a sign that our work has any merit. As we mentioned in the introduction of this chapter, we cannot combine vectors in two space and 3-space unless an adjustment is made first. Let $\vec{p} = 2\vec{i} + 5\vec{j}$ be entered in the form $L_4 = \{2, 5\}$. It clearly makes sense to form $\vec{u} + \vec{p}$, but notice that entering $L_1 + L_4$ on our calculator activates the error message DIM MISMATCH. To compute $\vec{u}+\vec{p}$, we must first enter $\vec{p} = \{2, 5, 0\} = L_4$, before the addition operation is performed.

The length or norm of a vector $\vec{v} = a\vec{i} + b\vec{j} + c\vec{k}$ is defined by

$$|\vec{v}| = \sqrt{a^2 + b^2 + c^2}.$$

We can use the listable property of lists to compute $|\vec{v}|$ as well. If $L_1 = \{a, b, c\}$ represents a vector of the form $\vec{v}$ entered onto our calculator, then entering $L_1 * L_1$ would produce the list $\{a^2, b^2, c^2\}$. By itself, this is not a mathematically "legitimate" vector operation, but using this with the $sum(\ )$ command, which sums the elements in a list, we can write

$$|\vec{v}| = \sqrt{a^2 + b^2 + c^2} = \sqrt{(sum(L_1 * L_1)} = \sqrt{(sum(L_1 \wedge 2)}.$$

Recall that the $sum(\ )$ command is item #5 in the MATH submenu of the LIST menu. (★86★ To compute the length (also called the norm) of a formal vector A, press [2ND] [VECTR] [F3] to select the MATH submenu, [F3] to select the *norm* command. Enter *norm A* and press [ENTER] to get the length of A. ★86★)

Example 9.2 *Find a unit vector in direction $\vec{v} = 13\vec{i} + 4\vec{j} - 9\vec{k}$. Find a vector of length 18 in direction $\vec{v}$.*

Often, the first step in solving a problem involving vectors is to turn a vector into a unit vector. This frequently makes the vector much easier to deal with. Scalar multiplication (by a positive scalar) does not change the direction of a vector, and so **scalar multiplying a vector by the reciprocal of its length will turn it into a unit vector** with the same direction. This is such a basic vector operation, that the concept should be completely understood before going any further.

From the home screen, enter $\vec{v}$ as $L_1 = \{13, 4, -9\}$. To compute the norm of $\vec{v}$, enter $\sqrt{(sum(L_1 \wedge 2))}$ and press [STO▶] [ALPHA] (L) [ENTER] to store $|\vec{v}| = 16.30950643$ as L. To compute a unit vector in direction $\vec{v}$, we form $\vec{u} = \frac{1}{|\vec{v}|}\vec{v}$. Enter L_1/L and press [STO▶] [2ND] [L2] [ENTER] to store

$$\vec{u} = \{.7970811413, .2452557358, -.5518254055\}$$

as L_2. Once we have a unit vector it is easy to get a vector of any length. This is one reason why unit vectors are important. (★86★ The command *unitV* can be used to turn a formal vector A into a unit vector. Press [2ND] [VECTR] [F3] to select the MATH submenu. Press [F2] to select the command *unitV*, enter *unitV A* and press [ENTER]. ★86★) We form the vector $\vec{w} = 18\vec{u}$ of length 18 by entering $18 * L_2$ to get

$$\vec{w} = \{14.34746054, 4.414603244, -9.9328573\}. \blacksquare$$

Entering a vector valued expression on our calculator must be done with care. We explain the process by entering the function $\vec{r}(x) = x^2\vec{i} + x^3\vec{j} + x^4\vec{k}$. To start with, we place our calculator in "Func" mode by selecting this item on the fourth line of the MODE menu. Enter $Y1 = X \wedge 2$, $Y2 = X \wedge 3$, $Y3 = X \wedge 4$ in the

9.1. ADDITION AND SCALAR MULTIPLICATION

Y= menu. **The disappointing result of entering** $\{Y1, Y2, Y3\}$ **on the home screen and storing the result to the list** $\text{L}R$ **is really quite predictable.** Each expression $Y1, Y2, Y3$ has a current value, depending on what value X has. We can enter X on the home screen to determine its current value. Suppose its current value is $X = 3$. In this case we would have stored the constant $\{Y1, Y2, Y3\} = \{9, 27, 81\}$ to the list $\text{L}R$, and changing the value of X after this assignment would have no effect on the value of $\text{L}R$. Clearly, we have not created a function.

What we need is a procedure to automatically update the value of $\text{L}R$, whenever the value of X is changed, and the section entitled **"Attaching Formulas to List Names"** on pages 11-7, 11-8, of the Ti-83 Guidebook covers such a technique. These two pages will provide additional background for this method, and reading this material is encouraged. (⋆86⋆ Read pages 162-166 of the TI-86 Guidebook. The technique is somewhat different on this machine. ⋆86⋆)

To create our vector valued function, we enter "$\{Y1, Y2, Y3\}$" and press [STO▶] [ALPHA] [R] [ENTER] to store the expression as $\text{L}R$. (⋆86⋆ A different expression must be entered. This topic is discussed below. ⋆86⋆) The quotation mark, obtained by pressing [ALPHA] ["], at the beginning and end of the expression is critical. As a suggestion, the easiest way to enter the entire expression is to press the pair [2ND] [ENTRY] repeatedly until $\{Y1, Y2, Y3\} \to R$ reappears on the active line (it's up there somewhere from our work above). Move the cursor to the ({) symbol at the left hand side of the formula and press [2ND] [INS] [ALPHA] ["] to insert a double quote. Move the cursor to the ($\to$) symbol and press [2ND] [INS] [ALPHA] ["] to finish the editing and press [ENTER].

Now that we have our vector valued function $\text{L}R$, we experiment a little to make sure it behaves in the right way. From the home screen, press [2] [STO▶] [X,T,θ,n] [ENTER] to give X a value of 2. Enter $\text{L}R$ and press [ENTER] to get $\{4, 8, 16\}$. Change the value of X to 3 and repeat this evaluation to get $\{9, 27, 81\}$. That should confirm that our expression is behaving like a vector valued function of X.

⋆⋆⋆⋆86⋆⋆⋆⋆

The above technique for **creating a vector valued function** by "attaching a formula to a list" is a technique that can **only be done to a list**. This is the reason why we cannot simply do all of our vector valued work through the VECTOR menu.

To create the same vector valued function $\vec{r} = R = R(x)$ that we created above (which will actually be a list), enter

$$Form("\{x \wedge 2, x \wedge 3, x \wedge 4\}", R)$$

and press [ENTER]. If the functions $y1(x) = x \wedge 2, y1(x) = x \wedge 3, y1(x) = x \wedge 4$, already appear in the $y(x) =$ menu, we could, of course, enter

$$Form("\{y1, y2, y3\}", R)$$

instead. The $Form(\ )$ command is in the LIST OPS submenu. Press [2ND] [LIST] [F5] [MORE] [MORE] [MORE] [F4]. The critical double quote symbol is in the STRING menu. Press [2ND] [STRNG] [F1]. As we have mentioned many times before, A list can be used in a calculation, by just entering the name of the list on the active line.

There is no prefix that must be attached to the list on this device. To evaluate this vector valued function at $x = 3$, press [3] [STO▶] [X-VAR] [ENTER], enter R and press [ENTER] to get $\{9, 27, 81\}$.

It is evident that we will be using both lists and vectors to represent vectors on this device. Fortunately, it is easy to convert a list to a vector and vice versa. Look for the commands $li \blacktriangleright vc$ and $vc \blacktriangleright li$ in the OPS submenu of both the LIST and VECTOR menus. If $A = [3, 5, 2]$, then enter $vc \blacktriangleright li\ A$ and press [ENTER] to get $\{3, 5, 2\}$. If $B = \{8, 2, 7\}$, then enter $li \blacktriangleright vc\ B$ and press [ENTER] to get $[8, 2, 7]$. Both calculations can be performed through either the LIST or VECTOR menu.

As a general rule, vectors should be entered formally as vectors, unless there is an expectation that vector valued expressions will be involved. In that case, lists should be used. The two data types cannot be mixed in a single computation. If both data types are present in a calculation, use a conversion command before they are brought together in a single computation.

★★★★ 86 ★★★★

The next example may appear to be pointless, but it involves a type of calculation that is frequently a part of other problems. Our initial temptation might be to solve the problem one component at a time. Sometimes this is necessary, but a awkward and tedious component by component calculation should certainly be avoided if at all possible.

Example 9.3 *Let $\vec{f}$ be the vector $\vec{f}(x) = x^2\vec{i} + (x-4)^2\vec{j} + (9-2x)\vec{k}$ and suppose that $\vec{p}(x) = \vec{f}(x) + \vec{C}$ for some unknown constant vector $\vec{C}$. Determine a $\vec{C}$ such that $\vec{p}(7) = 18\vec{i} - 33\vec{j} + 4\vec{k}$.*

In the Y= menu, enter $Y1 = X \wedge 2$, $Y2 = (X-4) \wedge 2$, and $Y3 = 9 - 2 * X$. On the home screen, we set up our vector valued function $\vec{f}(x)$ by entering "$\{Y1, Y2, Y3\}$" and pressing [STO▶] [ALPHA] ([F]) [ENTER] to store the expression as $_LF$. Enter $\{18, -33, 4\}$ and press [STO▶] [ALPHA] ([P]) [7] [ENTER] to store the vector $\vec{p7} = 18\vec{i} - 33\vec{j} + 4\vec{k}$ to $_LP7$ (a name). **We can use more elaborate names like this one for lists, but only for lists.** Since
$$\vec{p}(x) = \vec{f}(x) + \vec{C}$$
for all x, it follows that
$$\vec{p7} = \vec{p}(7) = \vec{f}(7) + \vec{C}$$
and so $\vec{C} = \vec{p7} - \vec{f}(7)$. Press [7] [STO▶] [X,T,θ,n] [ENTER] to give X a value of 7. By selecting our named vectors from the LIST menu, enter $_LP7 - _LF$, and press [STO▶] [ALPHA] ([C]) [ENTER] to store the value $\vec{C} = \{-31, -42, 9\} = -31\vec{i} - 42\vec{j} + 9\vec{k}$ to $_LC$. Now that we have a value for $\vec{C}$, we can define $\vec{p}(x)$ by entering it on the home screen in the form "$_LF + _LC$". Press [STO▶] [ALPHA] ([P]) [ENTER] to enter $\vec{p}(x)$ as $_LP$. **Notice the quotation marks in this formula. They are critical!** We should evaluate $_LP$ for a few values of X to make sure it behaves like the vector valued function we intended to create, but this verification is omitted. ■

9.2 The Dot and Cross Products

The two most interesting vector operations are the dot product and cross product. If $\vec{v}$ and $\vec{w}$ are the vectors

$$\vec{v} = v_1\vec{i} + v_2\vec{j} + v_3\vec{k}, \ \vec{w} = w_1\vec{i} + w_2\vec{j} + w_3\vec{k},$$

then the dot and cross product are defined by

$$\vec{v}\cdot\vec{w} = v_1w_1 + v_2w_2 + v_3w_3, \ \vec{v}\times\vec{w} = (v_2w_3 - w_2v_3)\vec{i} + (v_3w_1 - w_3v_1)\vec{j} + (v_1w_2 - w_1v_2)\vec{k}$$

The important geometric rule

$$\vec{v}\cdot\vec{w} = |\vec{v}||\vec{w}|\cos(\theta)$$

can be used to determine the angle, θ, between $\vec{v}$, and $\vec{w}$. There is a similar (equally important) rule

$$|\vec{v}\times\vec{w}| = |\vec{v}||\vec{w}|\sin(\theta)$$

for cross products, but it is generally not a good idea to use this rule to determine the angle, θ, between $\vec{v}$, and $\vec{w}$. Finding the angle, θ, in this manner, means writing this equation in the form $\sin(\theta) = p$ for θ, where $p \geq 0$ is the obvious product and quotient of lengths of vectors. **For every such p, however, there are two angles, θ, in the interval $0 \leq \theta \leq \pi$, which satisfy this equation.** Only one of these angles is an appropriate answer, and this is the root of the problem.

It is easy to compute the dot product of two vectors using the "listable" property of lists. If $\text{L}V$ and $\text{L}W$ represent two vectors $\vec{v}$ and $\vec{w}$ entered on our calculator, then if follows immediately from the above definition, that

$$\vec{v}\cdot\vec{w} = sum(\text{L}V * \text{L}V).$$

This is similar to the method used to compute the length of a vector. The identity

$$|\vec{v}| = \sqrt{\vec{v}\cdot\vec{v}}$$

states this elementary, but nonetheless, very useful relationship in vector algebra. (★86★ If V and W represent formal vectors on this device, press [2ND] [VECTR] [F3] to select the MATH submenu. Press [F4] to paste the $dot(\ )$ command on the active line. Finish by entering $dot(V, W)$ and press [ENTER] to get the dot product of the vectors $\vec{v} = V$ and $\vec{w} = W$. ★86★.)

To create a cross product command, we create a program. (★86★ Not surprisingly, the command $cross(\ )$ appears in the MATH submenu of the VECTOR menu. Enter $cross(V, W)$ and press [ENTER] to get the cross product of two formal vectors $\vec{v} = V$ and $\vec{w} = W$. Obviously, no program is required on this device. ★86★) The vectors $\vec{v}$ and $\vec{w}$ that are brought into the program must be given names in the program, and it is quite possible that these names could conflict with the original names. If the names $\text{L}A$ and $\text{L}B$ are used in the program, they would probably work well, until we tried to load vectors into the program which were also named $\text{L}A$ and $\text{L}B$, especially if they were loaded in reverse order. For this reason, **we use, in our** program, the two somewhat awkward names, $\text{L}VCT1$ and $\text{L}VCT2$, with the expectation that no one would used names like this in their own work. This makes the program somewhat tedious to enter from the keyboard, but it only has to be entered once. It may be worth recalling that pressing the pair [2ND] [ALPHA] locks the alpha keyboard mode so that the letters VCT can be entered more conveniently. Press [PRGM] ▶ ▶ [ENTER] to select the NEW submenu of the PRGM menu, and then enter the following program. (The last two lines remove vectors, which are no longer needed, from storage.)

PROGRAM: CROSS
:DISP "LEFT VECTOR"
:Input $\text{L}VCT1$
:Disp "RIGHT VECTOR"
:INPUT $\text{L}VCT2$
:$\{\text{L}VCT1(2) * \text{L}VCT2(3) - \text{L}VCT2(2) * \text{L}VCT1(3),$
$\text{L}VCT1(3) * \text{L}VCT2(1) - \text{L}VCT2(3) * \text{L}VCT1(1),$
$\text{L}VCT1(1) * \text{L}VCT2(2) - \text{L}VCT2(1) * \text{L}VCT1(2)\}$
:Disp "CROSS="
:Disp Ans
:DelVar($\text{L}VCT1$)
:DelVar($\text{L}VCT2$)

Let us perform a few sample calculations, using these commands. We let $\vec{v} = 2\vec{i} + 4\vec{j} - 3\vec{k}$ and $\vec{w} = 7\vec{i} - \vec{j} + 2\vec{k}$ by entering and storing $\{2, 4, -3\}$ and $\{7, -1, 2\}$ to $\text{L}V$ and $\text{L}W$ respectively. By extracting the $sum(\ )$ command from the MATH submenu of the LIST menu, enter $sum(\text{L}V * \text{L}W)$ to get $\vec{v} \cdot \vec{w} = 4$. Press [PRGM] and select CROSS in the EXEC submenu. At the prompts, enter $\text{L}V$ and $\text{L}W$ (in that order) and get $\vec{v} \times \vec{w} = \{5, -25, -30\} = 5\vec{i} - 25\vec{j} - 30\vec{k}$. These computations should also be done by hand, to verify that we have generated the correct answers. ∎

★★★★ 86 ★★★★

As we have seen, working with lists and vectors on this machine can be very different from equivalent work on the TI-83. These differences have been recorded in parenthetical remarks throughout the previous pages of this chapter. To continue to interupt the main story line with a large number of these parenthetical remarks would create highly unreadable text, **and so we will return to our standard practice of writing basically for the TI-83. It is hoped that our remarks have been sufficient, and that you will be able to translate the main story line into activity for this device.** In this way, TI-86 specific remarks can be reduced to only a few.

★★★★ 86 ★★★★

The algebraic rules governing the behavior of dot products and cross products are of practical importance for a variety of reasons. When a calculator is used, there is an additional reason. The rules can sometimes be used to greatly reduce the number of key strokes required to complete a computation. We can not prove these rules with our calculator, but we can show by example, that the rules seem to

9.2. THE DOT AND CROSS PRODUCTS

hold. In the next example, we attempt to make our case as convincing as possible, by using vectors with a wide mix of numerical values.

Example 9.4 *Verify that the* Distributive Law of Cross Product over Addition

$$\vec{u} \times (\vec{v} + \vec{w}) = \vec{u} \times \vec{v} + \vec{u} \times \vec{w}$$

holds for the vectors $\vec{u} = 17\vec{i} + \sqrt{2}\vec{j} - 8.24\vec{k}$, $\vec{v} = -e^3\vec{i} + 487.9\vec{j} + 0.53\vec{k}$, $\vec{w} = 4\vec{i} - \cos(12)\vec{j} - 93\vec{k}$. *In the process, notice that the right hand side of the distributive law requires many more key strokes to compute.*

We enter the vectors $\vec{u}$, $\vec{v}$, and $\vec{w}$ as $\text{L}U$, $\text{L}V$, and $\text{L}W$. Press $\boxed{\text{PRGM}}$, select CROSS from the EXEC submenu, and press $\boxed{\text{ENTER}}$. At the first prompt, enter $\text{L}U$, at the second prompt enter $\text{L}V + \text{L}W$ to get

$$\vec{u} \times (\vec{v} + \vec{w}) = \{4145.614038, -1457.465176, 8302.702867\}.$$

Most of the answer cannot be seen, because scrolling is not permitted. Actually, it is not necessary to look at the answer at this time, but if a view of the answer is desired, press $\boxed{\text{2ND}}$ $\boxed{\text{[ANS]}}$ $\boxed{\text{ENTER}}$ and then scrolling through the digits in the answer is enabled. Finally, for later reference, we store the answer to $\text{L}L$ ("L" for left hand side).

To compute the right hand side, we run the CROSS program again, entering $\text{L}U$ and $\text{L}V$ at the prompts and storing the answer to $\text{L}A$. Run the CROSS program one more time, entering $\text{L}U$ and $\text{L}W$ at the prompts and storing the answer to $\text{L}B$. Finally, we enter $\text{L}A + \text{L}B$ and store the answer to $\text{L}R$ ("R" for right hand side). The answer, which gives us a value for $\vec{u} \times \vec{v} + \vec{u} \times \vec{w}$ is omitted. It is impressive how many more key strokes to took to compute $\text{L}R$ compared to $\text{L}L$.

We can now compare the answers $\text{L}L$ and $\text{L}R$. We displayed the value of $\text{L}L$ above, but in practice, we might not want to express such long answers on paper. **How are we to compare these long answers without first expressing them on paper? We can only scroll through one answer at a time!** One approach is to enter $\text{L}L - \text{L}R$. The answer, which should be the zero vector, is $\{0, 0, 0\}$. ∎

We used a simple strategy here, which is worth repeating. It can be visually taxing to see that two complicated vectors are equal to each other, by just looking at them. We used an approach which was easier on our eyes. **To show that $\vec{L} = \vec{R}$, just show instead, that $\vec{L} - \vec{R} = \vec{0}$.**

Some of the familiar algebraic rules that hold for addition and multiplication of real numbers do not hold in vector algebra. **Cross product, for example, is not an associative operation**. In the next example we investigate this curious behavior in the cross product. Almost any combination of three vectors $\vec{u}$, $\vec{v}$, and $\vec{w}$, will show that $\vec{u} \times (\vec{v} \times \vec{w})$ is different from $(\vec{u} \times \vec{v}) \times \vec{w}$. (Some combinations satisfy associativity, but the chances are remote that a random choice would produce such a combination.)

Example 9.5 *Prove, by example, that the* Associative Law

$$\vec{u} \times (\vec{v} \times \vec{w}) = (\vec{u} \times \vec{v}) \times \vec{w}$$

<u>DOES NOT HOLD</u>.

A valid proof, of course, only requires one counter example, so such a task is well suited for our calculator. We choose our vectors randomly. Let $\vec{u} = 4\vec{i} - 8\vec{j} - 3\vec{k}$, $\vec{v} = -7\vec{i} - \vec{j} + 9\vec{k}$, $\vec{w} = \vec{i} - 11\vec{j} + 4\vec{k}$, and enter these three vectors as LU, LV, and LW. Press [PRGM], select and enter CROSS from the EXEC submenu, and enter LU and LV at the prompts. **We don't need to store the result.** Run CROSS again, at the first prompt, press [2ND] [ANS], and at the second prompt, enter LW. This gives $(\vec{u} \times \vec{v}) \times \vec{w} = \{-720, 240, 840\}$, which we store to LR.

Run the CROSS program, and enter LV and LW at the prompts. Again there is no need to store the intermediate result. Run CROSS again, enter LU at the first prompt, and press [2ND] [ANS] at the second prompt (are we sure Ans is still $Ans = \vec{v} \times \vec{w}$? Rest assured that it is!). This gives $\vec{u} \times (\vec{v} \times \vec{w}) = \{-513, -597, 908\}$, which we store to LR. The vectors LR and LL are clearly different, so the cross product operator is not associative.

What, then, can we say about these two triple cross products? Among other things, the vector $(\vec{u} \times \vec{v}) \times \vec{w}$ is perpendicular to $\vec{u} \times \vec{v}$, and so it is in the plane formed by $\vec{u}$ and $\vec{v}$ (if all of the vectors are based at the same point). This means that it is a linear combination of $\vec{u}$ and $\vec{v}$. Given vectors, $\vec{u}$, $\vec{v}$, and $\vec{w}$, We can use the SOLVE menu to find scalars a and b such that $(\vec{u} \times \vec{v}) \times \vec{w} = a\vec{u} + b\vec{v}$. This and other interesting properties of this triple cross product will be explored in the exercises.

The operations we have been discussing in the last two sections can be used effectively in a large variety of mathematical problems. A sample of these standard mathematical problems are considered in the next section. The operations can also be used to solve very interesting and impressive applied problems. Some of these appear in the exercises at the end of this chapter. In particular, Captain Ralph reappears after a long vacation. Now that we are working in three dimensional space we will have the opportunity to accompany Captain Ralph more frequently on his adventures.

We end this section with one standard mathematical application. Frequently, vectors need to be decomposed into a sum of two vectors, one which is parallel to a given direction, and the other which is perpendicular. For example, if $\vec{f}$ is a force vector being applied to an object traveling along a curve C, then it would make sense to express $\vec{f}$ in the form

$$\vec{f} = \vec{f}_t + \vec{f}_n,$$

where $\vec{f}_t$ is tangent to the curve C, and $\vec{f}_n$ is normal to the curve C.

Example 9.6 *Express the vector $\vec{v} = 13\vec{i} - 32\vec{j} + 7\vec{k}$ as the sum of a vector $\vec{w}_1$ parallel to $\vec{w} = -8\vec{i} + 9\vec{j} + 2\vec{k}$, and a vector $\vec{w}_2$ which is perpendicular to $\vec{w}$.*

Before we start, **we remove some of the clutter on our calculator** by pressing [2ND] [MEM] [2] (for delete) [4] (for lists). Press [ENTER] wherever the cursor is located to delete that list from memory.

It helps to keep a picture in mind to motivate our work. There are two different figures shown corresponding to two substantially different circumstances, one where the angle θ between $\vec{v}$ and $\vec{w}$ is acute and one where it is obtuse.

9.2. THE DOT AND CROSS PRODUCTS

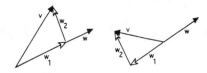

We project $\vec{v}$ onto $\vec{w}$ to get $\vec{w}_1$. Once we have $\vec{w}_1$, the vector $\vec{w}_2$ is easy to get from $\vec{v} = \vec{w}_1 + \vec{w}_2$ by subtraction.

Calculus texts always have a formula for the projection vector, but there is no need to memorize this formula. Just look at the picture. We first turn $\vec{w}$ into a unit vector $\vec{u}$—this is the direction of $\vec{w}_1$, or its opposite direction as we can see from the figure. Then $\vec{w}_1 = L\vec{u}$, where L is either the length of the projection vector, or its negative depending on whether $\vec{w}_1$ and $\vec{u}$ have the same or opposite directions. Elementary trigonometry easily implies that $L = |\vec{v}|\cos(\theta)$, where θ is the angle between $\vec{v}$ and $\vec{u}$. This number L automatically has the right sign as well as the right size.

$$L = |\vec{v}|\cos(\theta) = |\vec{v}|\frac{\vec{v}\cdot\vec{w}}{|\vec{v}||\vec{w}|} = \frac{\vec{v}\cdot\vec{w}}{|\vec{w}|}.$$

We enter $\vec{v}$ and $\vec{w}$ as LV and LW. We can then enter

$$sum(\text{L}V * \text{L}W)/\sqrt{(sum(\text{L}W \wedge 2))}$$

and store the result as $L = -30.9669666$. To compute $\vec{w}_1 = L\frac{1}{|\vec{w}|}\vec{w}$, we enter

$$\text{L}W * L/\sqrt{(sum(\text{L}W \wedge 2))}$$

and store the result as

$$\vec{w}_1 = W1 = \{20.29530201, -22.83221477, -5.073825503\}.$$

Enter LV − LW1 and store the result as

$$\vec{w}_2 = W2 = \{-7.295302013, -9.167785235, 12.0738255\}.$$

We check or answers by computing LW1 + LW2 to get $\vec{w}_1 + \vec{w}_2 = \{13, -32, 7\}$, which is indeed $\vec{v}$. We check the orthogonality of the vectors $\vec{w}_2$ and $\vec{w}$ by computing $sum(\text{L}W2 * \text{L}W)$ to get $\vec{w}_2 \cdot \vec{w} = -2.1\,10^{-11} \approx 0$, which (allowing for a small round off error) implies orthogonality.

We're done, but as an added touch, we include a conversion over to a rational form for these answers. We have used this command before, but not for some time, so **this is also a reminder of an available option.** Enter LW1 and press MATH ▶ 1 to select ▶Frac. Press ENTER to get

$$\vec{w}_1 = \{\frac{3024}{149}, \frac{-3402}{149}, \frac{-756}{149}\}.$$

In the same way, we get

$$\vec{w}_2 = \{\frac{-1087}{149}, \frac{-1366}{149}, \frac{1799}{149}\}.$$

The decimal numbers from which these values are derived have probable round off errors, and so one would expect these rational forms to be slightly less than perfect. It turns out, however, that the above rational values for $\vec{w}_1$ and $\vec{w}_2$ are exact. ■

9.3 Lines and Planes

Example 9.7 *Find the equation of the plane which passes through the points* $P(2, -5, -4)$, $Q(-1, 6, 7)$, $R(9, 5, -6)$.

Mathematically, a point and a vector in three dimensional space are equivalent. They are both just ordered triplets of numbers. The only difference is a point of view. While this is frequently an advantage, it can also cause confusion. On paper, we will use notation to separate the two ideas. With some practice, we will learn to flow from one notion to the other smoothly and without fanfare. We will use notation such as P and $\vec{P}$ or $\vec{p}$ on paper to denote an ordered triplet of numbers thought of as a point and as a vector, respectively, but once this triplet is entered on our calculator, it will simply become LP for both points of view.

We begin by entering the points P, Q, R, as lists (points) (vectors) LP, LQ, LR, on our calculator. The problem of finding the equation of a plane almost always involves a search for a normal (perpendicular) vector to the plane. In the present case, the vectors $\vec{PQ}$ and $\vec{PR}$ from P to Q and from P to R both lie in the plane, and so the vectorindexvector(s)!normal $\vec{n} = \vec{PQ} \times \vec{PR}$, which is normal to both $\vec{PQ}$ and $\vec{PR}$ is normal to the plane as well. To enter $\vec{n}$, press PRGM and select CROSS from the EXEC submenu. At the first prompt enter LQ − LP, and at the second prompt enter LR − LP. Store the result as LN to get

$$\vec{n} = \text{L}N = \{-132, 71, -107\}.$$

The point $Z(x, y, z)$ is in the plane if and only if the vector from P to Z is orthogonal to the normal vector $\vec{n}$. This means that

$$\vec{n} \cdot \vec{PZ} = 0,$$

which gives us the equation of the plane. If $\vec{n} = \{n_1, n_2, n_3\}$ and $P = (p_1, p_2, p_3)$, notice that this equation can be expressed as

$$n_1(x - p_1) + n_2(y - p_2) + n_3(z - p_3) = 0,$$

or, in other words as

$$n_1 x + n_2 y + n_3 z = n_1 p_1 + n_2 p_2 + n_3 p_3 = \vec{n} \cdot \vec{p}.$$

The right hand side of this equation is easily computed on our calculator by entering $sum(\text{L}N * \text{L}P)$ to get a value of −191. The equation of the plane is

$$-132x + 71y - 107z = -191. \qquad \blacksquare$$

Example 9.8 *Find the distance between the line through* $A(4, -3, 8)$, $B(-1, 9, 5)$ *and the line through* $C(5, 7, -2)$, *and* $D(13, 2, -6)$.

Imagine a plane which contains one of the lines and is parallel to the second line. Then the distance in question is just the distance from a point (any point on the second line) to the plane. Most calculus books provide a formula for the distance

9.3. LINES AND PLANES

from a point Q to a plane, but it is fairly easy to see geometrically. Just take any vector from a point A in the plane to Q, project it onto the plane's normal vector $\vec{n}$, and compute the length of the projection of $\vec{AC}$ onto $\vec{n}$. If this strategy seems mysterious, just draw a picture to clear the air. Look at the plane on edge, draw the normal vector $\vec{n}$ based at the point A on the plane, draw the point Q off of the plane, followed by $\vec{AQ}$ and the projection of $\vec{AQ}$ onto $\vec{n}$.

We enter the points A, B, C, D on our calculator as LA, LB, LC, LD. The vectors $\vec{AB}$ and $\vec{CD}$ are both parallel to the plane described above, and so $\vec{n} = \vec{AB} \times \vec{CD}$ is normal to the plane. To enter $\vec{n}$ on our calculator, we press [PRGM] and select the CROSS program. Enter LB − LA at the first prompt, and LD − LC at the second prompt. Store the result as LN to get the normal vector

$$\vec{n} = \text{L}N = \{-63, -44, -71\}.$$

For the sake of argument, suppose the plane contains the line through A and B. To find the distance, we find the distance between C (or any other point on the line through C and D) and the plane. We don't need the equation of the plane to compute this. We simply project the vector from A to C onto $\vec{n}$ and compute its length. In the distance formula below, $\cos(\theta)$ could be positive or negative depending on whether $\vec{n}$ and C are on the same or opposite sides of the plane. To make the distance positive, we use $|\cos(\theta)|$ instead of $\cos(\theta)$.

$$d = |\vec{AC}||\cos(\theta)| = |\vec{AC}|\left|\frac{\vec{AC}\cdot\vec{n}}{|\vec{AC}||\vec{n}|}\right| = \left|\frac{\vec{AC}\cdot\vec{n}}{|\vec{n}|}\right|$$

On our calculator we enter

$$abs(sum((\text{L}C - \text{L}A) * \text{L}N))/\sqrt{(sum(\text{L}N \wedge 2))}$$

to get a distance of $d = 1.978529926$. ∎

We could have used any point in the plane instead of A and any point on the line through C and D instead of C. It is a worthwhile and interesting exercise to go through these calculations again using B instead of A, and D instead of C.

There is a formula for the distance between two lines in space, just as there is a formula for the distance between a point and a plane. We have appealed to neither of these formulas. Drawing pictures and formulating these computations each time they are used is a positive learning experience. Besides, formulas, empty of motivation, are apt to be forgotten as soon as they are not used.

This problem can also be done by minimizing a certain function of two variables, namely the function giving the distance between a variable point on one line and a variable point on the other.

Example 9.9 *Find an equation for the line of intersection between the plane through $A(5, 1, -8)$, $B(-6, 2, 7)$, $C(3, 9, 1)$, and the plane through $P(-8, 1, 1)$, $Q(4, -7, -3)$, and $R(-5, -7, -1)$.*

There are two basic ways of proceeding. In one approach, the equation of the line is found by solving a system of two equations in three unknowns. This problem is a part of a more general problem involving m equations in n unknowns, and it is one of the main topics of discussion in a linear algebra course usually taken soon after a calculus sequence.

The approach we take is more geometrical. Let $\vec{n_1}$ and $\vec{n_2}$ be normal vectors for each of the two planes. The line of intersection lies in both planes and so it is orthogonal to both $\vec{n_1}$, and $\vec{n_2}$. It follows that the vector $\vec{v} = \vec{n_1} \times \vec{n_2}$ is a direction vector for the line of intersection. All we need, then, is a point of intersection of the two lines.

Our calculator is accumulating many vectors which are no longer being used, so let us begin by pressing [2ND] [MEM] [2] [4], and then pressing [ENTER] to delete each list that is no longer needed.

Enter A, B, C, P, Q, R as ᴌA, ᴌB, ᴌC, ᴌP, ᴌQ, ᴌR. The normal vector $\vec{n_1}$ to the plane ABC through the points A, B, C, can be expressed in the form $\vec{n_1} = \vec{AB} \times \vec{AC}$. The normal vector $\vec{n_2}$ to the plane PQR through the points P, Q, R, can be expressed in the form $\vec{n_2} = \vec{PQ} \times \vec{PR}$. Press [PRGM] and select CROSS. Enter ᴌB − ᴌA at the first prompt, and ᴌC − ᴌA at the second. Store the result as ᴌN1 to get

$$\vec{n_1} = \text{ᴌN1} = \{-111, 69, -86\}.$$

Running the CROSS program again, we enter ᴌQ − ᴌP at the first prompt, and ᴌR − ᴌP at the second. Store the result as ᴌN2 to get

$$\vec{n_2} = \text{ᴌN2} = \{-16, 12, -72\}.$$

To get the direction vector $\vec{v} = \vec{n_1} \times \vec{n_2}$ for the line, we run the CROSS program again, entering ᴌN1 at the first prompt and ᴌN2 at the second. Store the result as ᴌV, and we have our direction vector

$$\vec{v} = \text{ᴌV} = \{-3936, -6616, -228\}.$$

The next step is "window dressing", but rather interesting window dressing. The vector $\vec{v}$ has rather large components, and we can replace it by any scalar multiple (divisor) designed to bring down its size. Of course, if the resulting vector no longer has integer components, then it is not a simpler vector. We wish to divide ᴌV by the largest integer d, which appears as a factor in each of the elements in ᴌV. Such a number is called the *greatest common divisor* or *GCD* of the elements in the list ᴌV. Our calculator has such a command, but it can only be used to compute the GCD of pairs of numbers (which must be nonnegative). Press [MATH] ▶ to select the NUM submenu, and press [9] to select $gcd(\ )$. If a and b are positive integers, $gcd(a, b)$ returns the greatest common divisor of a and b. An entry of the form $gcd(a, \{b_1, b_2, \ldots, b_j\})$ returns a list $\{c_1, c_2, \ldots, c_j\}$ where each $c_k = gcd(a, b_k)$. As you can see, whether individual integers or lists are used, the computation is always done in pairs.

To compute the GCD of our list ᴌV with three elements, we enter $gcd(6616, 228)$ and press [ENTER] to get 4. Then enter $gcd(3936, \{6616, 228\})$ to get $\{8, 12\}$. It follows that the GCD of the triplet is 4.

We are ready to bring down the size of $\vec{v}$. At the same time, we may as well alter the vector by a negative sign—the result will certainly be another valid direction

9.3. LINES AND PLANES

vector. Enter $\text{L}V/(-4)$, and press [STO▶] [ALPHA] ([v]) [ENTER] to change the value of $\text{L}V$ to the direction vector
$$\vec{v} = \text{L}V = \{984, 1654, 57\}.$$

To specify the line, we need one point on the line, and for this we will need the equations of the planes. If $Z(x, y, z)$ represents a point on either plane, then the vector equations
$$\vec{n_1} \cdot \vec{AZ} = 0, \ \vec{n_2} \cdot \vec{PZ} = 0$$

represent the equations of the planes. Just like in Example 9.7, these equations can be expanded and expressed in the form
$$\vec{n_1} \cdot \vec{Z} = \vec{n_1} \cdot \text{L}A, \vec{n_2} \cdot \vec{Z} = \vec{n_2} \cdot \text{L}P,$$

which lends itself to a convenient calculator simplification. We enter $sum(\text{L}N1 * \text{L}A)$ to get 202. Enter $sum(\text{L}N2 * \text{L}P)$ to get 68. With this we can write the equations of the planes as
$$-111x + 69y - 86z = 202, \ -16x + 12y - 72z = 68.$$

We only need one point, so we could specify a value for one of the variables and solve the two equations and two unknowns for the remaining variables. We set $z = 0$.

The "proper way" to solve this system of two linear equations in two unknowns on our calculator is to use **linear algebra and matrix theory**. This, however, would take us too far astray from our current interests, and so we use instead another very natural approach. With $z = 0$ solve the second equation by hand to get $y = (68 + 16x)/12$. Press [Y=] and enter $Y1 = (68 + 16 * X)/12$. Press [MATH] [0] to bring up the EQUATION SOLVER screen, and enter the first equation in the form
$$eqn : 0 = -111 * X + 69 * Y1 - 202.$$

Recall that $Y1$ is entered by pressing [VARS] [▶] [1] [1]. Press ▼ to make the $X =$ line the active line and press [ALPHA] ([SOLVE]). (The solution is unique and numerically uncomplicated, so there is no need to search for a start up value first.) This gives us a value of $x = X = 9.9473684210$ and by pressing [VARS] [1] [1], a value of $y = Y1 = 18.92982456$. The point
$$S = S(x_0, y_0, z_0) = (9.9473684210, 18.92982456, 0)$$

is a point on the line. If $Z(x, y, z)$ represents an arbitrary point on the line of intersection, then the idea that $\vec{SZ}$ is a scalar multiple of $\vec{v}$ leads to the vector valued equation
$$\vec{SZ} = t\vec{v},$$

as the equation of the line. Expanding this, we get the parametric form

$$x = 9.9473684210 + 984t, \ y = 18.92982456 + 1654t, \ z = 0 + 57t. \qquad \blacksquare$$

CHAPTER 9. VECTORS AND ANALYTIC GEOMETRY

Example 9.10 *Captain Ralph, on his way back to his home star base, after a routine patrol on the dark side of the moon, hands control of his star ship over to his able copilot Lundar. Too bored to stare at the emptiness of space any longer, he thinks about taking a practice shot at an old spherical communication satellite, 14 meters in diameter, that is passing by. Lundar says that it is no longer functional, but Ralph is not so sure. Nevertheless, when the ship's laser gun is at position*

$$P(142.875, -813.412, 293.843),$$

Captain Ralph points the gun in the direction of the vector

$$\vec{u} = -0.889\vec{i} + 0.381\vec{j} + 0.254\vec{k},$$

and unable to resist the temptation any longer, he pulls the trigger. Does he hit the satellite, which is (at that moment) centered at position

$$Q(69.37547, -782.1291, 314.74373)?$$

If not, by how much does he miss. Units are in kilometers in a coordinate system having its origin at the home star base.

In the usual way, we enter P, Q, and $\vec{u}$ on our calculator as LP, LQ, and LU. Let $Z(x, y, z)$ denote an arbitrary point on the straight line formed by the laser beam. Using the idea that the vector $\vec{PZ}$ is a scalar multiple of $\vec{u}$, we write the equation of the line as $\vec{PZ} = t\vec{u}$, which can then be expressed in the form $\vec{Z} = \vec{P} + t\vec{u}$. While it is not an important matter, we put our calculator in parametric mode for no other reason other than to produce a T by pressing the $\boxed{\text{X,T,}\theta\text{,}n}$ key. **Notice how easy it is to enter the equation of the line on our calculator.** Simply enter "L$P + T *$ LU", and store the result as LR. We have discussed the importance of the double quotes in our previous examples. With LR defined in this way, its values will be updated whenever we change the value of T.

Actually, a formula for this line is not needed. All we really need is the distance between the line and Q. A picture will help.

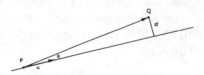

The picture, in fact, helps enormously. Using elementary right triangle trigonometry, we can see that the distance

$$d = |\vec{PQ}|\sin(\theta) = |\vec{PQ}|\frac{|\vec{PQ} \times \vec{u}|}{|\vec{PQ}||\vec{u}|} = \frac{|\vec{PQ} \times \vec{u}|}{|\vec{u}|}.$$

We compute the cross product first. Press $\boxed{\text{PRGM}}$, select CROSS, enter L$Q -$ LP at the first prompt, LU at the second, and store the result to LV. Then compute

$$\sqrt{(sum(\text{L}V \wedge 2)/sum(\text{L}U \wedge 2))}$$

9.4. EXERCISE SET

to determine $d = .0075677357$ kilometers, or in other words $d = 7.5677357$ meters. Nice going Captain Ralph. You missed by 0.5677357 meters! ■

Projection vectors were used frequently throughout this section, and we always went back to basic vector work to construct the projections. This is a good learning device, but eventually, it may be more efficient to create a program which computes the projection vector. This is a problem in the exercise set. Once such a program is created, it can be used to easily compute projection vectors.

9.4 Exercise Set

1. Let $\vec{u}, \vec{v}, \vec{w}$ be the vectors defined by $\vec{u} = 7\vec{i} - 11\vec{j} - 9\vec{k}$, $\vec{v} = -13\vec{i} + 5\vec{j} - \vec{k}$, and $\vec{w} = 2\vec{i} + 3\vec{j} + 6\vec{k}$. Compute the following vectors. Some parts have obvious answers. Try to catch them in advance, and state the answer before you use your calculator. Why are parentheses not used in parts (m) and (n)?

   a) $\vec{u} + \vec{v}$
   b) $\vec{w} - \vec{v}$
   c) $3\vec{u} + 7\vec{w}$
   d) $7\vec{v} - 12\vec{w}$
   e) $\vec{u} + \vec{v} + \vec{w}$
   f) $5\vec{u} + 8\vec{v} - 3\vec{w}$
   g) $\vec{u} \cdot \vec{u}$
   h) $\vec{u} \times \vec{v}$
   i) $\vec{u} \times (\vec{v} \times \vec{w})$
   j) $(\vec{u} \times \vec{v}) \times \vec{w}$
   k) $(\vec{u} \times \vec{v}) \times \vec{v}$
   l) $\vec{w} \times (13\vec{w})$
   m) $\vec{u} \cdot \vec{v} \times \vec{w}$
   n) $\vec{v} \cdot \vec{v} \times \vec{w}$
   o) $(3\vec{u} + 7\vec{w}) \cdot (7\vec{v} - 12\vec{w})$

2. Let $\vec{u}$, and $\vec{v}$ be the vectors $\vec{u} = 8\vec{i} - 3\vec{j} - 6\vec{k}$, $\vec{v} = 2\vec{i} + 5\vec{j} + 7\vec{k}$.
   a) Compute the lengths of $\vec{u}$, and $\vec{v}$.
   b) Compute the length of $5\vec{u} - 8\vec{v}$.
   c) Compute the length of $\vec{u} \times \vec{v}$.
   d) Find a unit vector in the direction of $2\vec{u} - 3\vec{v}$.
   e) Find a vector of length 7 in the direction of $\vec{u} \times \vec{v}$.
   f) Find the angle between $\vec{u}$, and $\vec{v}$ in degree measure.
   g) Find the projection of $\vec{v}$ onto $\vec{u}$.
   h) Express $\vec{v}$ in the form $\vec{v} = \vec{v_1} + \vec{v_2}$, where $\vec{v_1}$ is parallel to $\vec{u}$, and $\vec{v_2}$ is orthogonal to $\vec{u}$.

3. Create a projection program (call it PROJ), so that if $_LU$, and $_LV$ are names for the vectors $\vec{u}$, and $\vec{v}$, then PROJ computes the projection of $\vec{u}$ onto $\vec{v}$. Use it to compute the projection of $\vec{a} = 6\vec{i} - \vec{j} - 613\vec{k}$, onto $\vec{b} = 2\vec{i} + 11\vec{j} - 5\vec{k}$, and the projection of $\vec{b}$ onto $\vec{a}$.

4. Verify that $\vec{u}$, and $\vec{v}$ are orthogonal to $\vec{u} \times \vec{v}$, for $\vec{u} = 35\vec{i} + 87\vec{j} - 14\vec{k}$ and $\vec{v} = -19\vec{i} + 4\vec{j} + 286\vec{k}$

5. Verify that the *Distributive Law of Dot Product Over Addition* holds for a range of values of $\vec{u}, \vec{v}$, and $\vec{w}$. More important, Explain how to do this on your calculator with a minimum of key strokes. Think about using the double quote symbol ("), and perhaps the [2ND] [ENTRY] and [2ND] [RCL] keys.

6. Verify that the identity $|\vec{u} \times \vec{v}|^2 = |\vec{u}|^2|\vec{v}|^2 - (\vec{u} \cdot \vec{v})^2$ holds for $\vec{u} = 57\vec{i} + 4.3\vec{j} - 23.8\vec{k}$ and $\vec{v} = -196\vec{i} + 4\vec{j} + 28\vec{k}$. (The proof in general is a straight forward but very tedious calculation. One lets $\vec{u} = \{u_1, u_2, u_3\}$, $\vec{v} = \{v_1, v_2, v_3\}$, $\vec{w} = \{w_1, w_2, w_3\}$,

and simply computes both sides.) This algebraic result is important, because it leads immediately to a proof of the fundamental geometric rule concerning the length of a cross product, namely,

$$|\vec{u} \times \vec{v}| = |\vec{u}||\vec{v}|\sin(\theta)$$

where θ $(0 \leq \theta \leq 2\pi)$ is the angle between $\vec{u}$, and $\vec{v}$. Explain why this follows immediately.

7. For reasons of orthogonality that were discussed on page 160, it follows that $(\vec{u} \times \vec{v}) \times \vec{w}$ is a linear combination of $\vec{u}$, and $\vec{v}$. For $\vec{u} = 82\vec{i} + 13\vec{j} - 37\vec{k}$, $\vec{v} = -19\vec{i} + 4\vec{j} + 286\vec{k}$, $\vec{w} = \vec{i} - 6\vec{j} + 38\vec{k}$, find scalars a, and b, such that $(\vec{u} \times \vec{v}) \times \vec{w} = a\vec{u} + b\vec{v}$.

8 Verify the rule $(\vec{u} \times \vec{v}) \times \vec{w} = \vec{u} \times (\vec{v} \times \vec{w}) + (\vec{u} \times \vec{w}) \times \vec{v}$ for the vectors $\vec{u} = 35\vec{i} + 87\vec{j} - 14\vec{k}$, $\vec{v} = -19\vec{i} + 4\vec{j} + 286\vec{k}$, $\vec{w} = \vec{i} - 6\vec{j} + 38\vec{k}$. Use this rule to create a nontrivial example of vectors $\vec{u}$, $\vec{v}$, and $\vec{w}$ for which $(\vec{u} \times \vec{v}) \times \vec{w} = \vec{u} \times (\vec{v} \times \vec{w})$. (By nontrivial, we mean that the final products should be nonzero.)

9. Consider the parallelepiped formed by the three vectors $\vec{u} = 2\vec{i} - 7\vec{j} + 5\vec{k}$, $\vec{v} = 4\vec{i} + 2\vec{j} + 5\vec{k}$, $\vec{w} = 6\vec{i} + 9\vec{j} - 3\vec{k}$ based at a common vertex P. Determine the angle (in degree measure) between the main diagonal from P (to the opposite corner of the parallelepiped), and the diagonal from P, along each of the three faces adjacent to P.

10. Let $\vec{u}$, $\vec{v}$, and $\vec{w}$, be any three noncolinear vectors in three dimensional space, Let R be the parallelepiped formed by $\vec{u}$, $\vec{v}$, $\vec{w}$ based at a common vertex P. Let $\vec{f1}$, $\vec{f2}$, and $\vec{f3}$ be the diagonal vectors from P along the three faces of R adjacent to P, and finally, let Rf be the parallelepiped formed by $\vec{f1}$, $\vec{f2}$, and $\vec{f3}$. It turns out that the volume of Rf is always exactly twice the volume of R. Verify this result for the case $\vec{u} = 35\vec{i} + 87\vec{j} - 14\vec{k}$, $\vec{v} = -19\vec{i} + 4\vec{j} + 286\vec{k}$, $\vec{w} = \vec{i} - 6\vec{j} + 38\vec{k}$. (This can also be shown by hand—in just a few lines—using the algebraic properties of the vector operations.)

11. Enter an expression on your calculator which computes the points on the line through the points $P(-12, 18, 25)$ and $Q(38, -29, 13)$ This should be done in such a way that points on the line are easily generated. Explain the procedure you used. Compute a few sample values of points on the line.

12 Find the equation of the line through the point $P(27.94, 13.37, -43.12)$, which passes 3 units away from the point $Q(83.15, 92, 87, 17.02)$.

13. Find the equation of the line through $A(17, -42, 31)$, which intersects at right angles, the line through $B(-7, 12, -22)$ with direction vector $\vec{v} = 3\vec{i} - 6\vec{j} - 7\vec{k}$. Find the point of intersection of the two lines.

14. Find the equation of the plane through the points $A(5, -9, 1)$, $B(16, -2, 31)$, $C(-8, 53, -17)$.

15. Find the equation of the plane which contains the point $A(7, 15, -8)$ and the line defined parametrically by $x = 4 + 2t, y = -6 + 5t, z = 7 - 3t$.

16. Find the equation of the plane which contains the point $A(-4, 9, 6)$, is parallel to, and 2 units away from, the line L through the points $B(7, 5, -3)$, and $C(-6, 4, 13)$.

9.4. EXERCISE SET

17. Find the parametric equation of the line of intersection between the plane passing through $A(2, 1, -5)$, $B(-4, 11, 8)$, $C(12, 3, -2)$ and the plane passing through $P(14, 23, 18)$, $Q(-9, -1, -32)$, $R(-1, 19, -6)$.

18. How would you define the angle between a plane and a nonparallel line? Define this term. Find the angle between the plane whose equation is $5x - 8y + 2z = 17$, and the line with equation $x = 7 - t, y = 2 + 9t, z = -6 + 5t$.

19. Create a program which computes the distance between two lines. Your program should call for input information A, $\vec{u}$, B, $\vec{v}$, and the output should be the distance between the line through A with direction $\vec{u}$, and the line through B with direction $\vec{v}$. Use it to compute the distance between the line through $A(8, 1, -4)$, $B(-2, 5, 3)$ and the line through $P(14, 5, -6)$, $Q(-3.-17, 2)$.

20. Captain Ralph has spotted a small but deadly object at position

$$P(232.627, 632.879, 3275.68)$$

traveling at great speed in a straight line with direction vector

$$\vec{v} = 1.2764\vec{i} - 0.26271\vec{j} + 1.7014\vec{k}$$

towards a spherical space station of radius 250 meters centered at

$$Q([258.257, 627.413, 3309.599]).$$

Will it hit the space station, and if not, how close will it come? The coordinate system is in kilometers.

21. A damaged space ship from United Earth Federation (UEF) is hiding from the dreaded Lizard Warriors, and attempting to make radio contact with its headquarters without being observed by the enemy. A UEF observer at position $P(73.1, -19.8, -42.6)$ has determined that a brief radio signal was broadcast from somewhere in the direction of the vector $\vec{u} = 3.2\vec{i} + 1.2\vec{j} - 5.1\vec{k}$ based at P. At the same time, another UEF observer at position $Q(-1.4, 81.9, -13.7)$ received a radio signal broadcast from somewhere in the direction of the vector $\vec{v} = 9.4\vec{i} - 3.0\vec{j} - 10.5\vec{k}$ based at Q. The coordinate system is expressed in kilometers. Determine the position of the damaged ship. There are observable errors in these measurements. Account for them.

Captain Ralph will launch a fast rescue mission only if the location data is accurate enough to justify the risk. The two direction lines used to locate the damaged ship should meet, but as long as they are less than 1 kilometer apart, the data will be considered reliable enough to begin the operation. Captain Ralph, always ready to spring into action, awaits your call. Should he rescue, or should he not?

Project: Captain Ralph Shoots at a Mirror

A planar circular mirror with a radius of 10 meters is built, and great effort is taken to make the mirror absolutely flat. The mirror is then mounted to the ends of three, parallel, variable length rods which project from the outer wall of a space station. A heavy, fixed, mounting device from the wall of the space station is attached to the center of the mirror's underside, which secures the center of the mirror in a fixed position, and the mirror is then allowed to pivot on its center, in a way which is determined by the lengths of the three adjustment rods. Since the center of the mirror is fixed, the lengths of two of the adjustable rods determine a unique setting of the mirror, and the length of the third rod is then adjusted to provide a secure position for the mirror.

The *mirror's target vector* is defined to be the unit vector, normal to the planar surface of the mirror, which, if based at the mirror's center, would point away from the space station.

The mirror must be mounted so that its target vector is exactly the same as the unit vector with direction

$$\vec{n} = \sqrt{3}\vec{i} + \sqrt{7}\vec{j} + 42\sqrt{2}\vec{k}$$

(relative to a certain three dimensional coordinate system discussed below), and the lengths of the rods are to be adjusted to product this direction. The adjustment must be made with great accuracy.

Since Captain Ralph happens to be in the right region, on his way back to the space station for lunch, he has been asked to help out. He is told to maneuver his space ship to a position which is roughly in line with the mirror's target vector, and about 30 kilometers away. Here he is to shoot a beam of light from a laser gun in his ship at the mirror, adjust his position and shoot again, and continue doing this until the beam is reflected by the mirror and returned to exactly its original position at the gun. A record will then be made of the gun's coordinate position at this time, and this information will be used to adjust the mirror's position.

Being the best pilot in the fleet, Captain Ralph does the job in short order. While maneuvering his ship, he takes a test shot here and there, and soon sees the tell tale reflection of the laser beam appearing back in the gun's sight. As instructed, he marks his position (actually, the gun's position) as

$$P(873.57838, 1334.4355, 29957.581).$$

Here is some information concerning the rods and the coordinate system. All units are measured in *meters*. The coordinate system is arranged so that the fixed center of the mirror is at (0,0,0). The z-axis is parallel to the adjustment rods with a positive direction pointing away from the space station. The positive x and y axes are shown in the figure, along with cross sections of the three rods which are labeled a, b, and c. The adjustment rods are 9.5 *meters* away from the z-axis, and are equally distributed around the z-axis as suggested in the picture.

1. Exactly where the laser light beam strikes the mirror is unknown, and so the center of the mirror is used as the point where the light beam strikes. What should the adjustment be at each adjustment rod in order to produce a mirror position whose target vector has direction $\vec{n}$? How many centimeters should each rod be lengthened or shortened?

9.4. EXERCISE SET

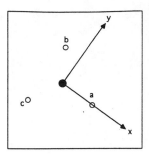

Remember that the coordinate system is in meters, and the adjustment specifications are to be made in centimeters. Good luck. This has to be accurate!

2. The laser light beam could, of course, strike any point on the mirror, not necessarily its center, and this could cause an error in positioning the mirror. Determine the size of the error in the following way. If the final setting of the mirror results in a target vector $\vec{n_0}$, then determine, in degree measure, the largest possible angle between $\vec{n_0}$, and the vector $\vec{n}$.

Chapter 10

Vector Valued Functions

Vector valued functions of a real variable t play an important roll in mathematics and its applications. Now that we have broken out of a one dimensional world, we can look forward to a variety of really interesting problems. If done by hand, calculations can quickly get out of hand, but fortunately, we can look forward to a big help from our calculator.

10.1 Basic Calculations

We discussed how to enter a vector valued function of a real variable on our calculator in the last chapter, but this is a process worth reviewing. To enter the function

$$\vec{f}(x) = (5x^3 - x)\vec{i} + (2x^2 + 4)\vec{j} + (2x - 8)\vec{k},$$

we press [Y=] and enter $Y1 = 5*X \wedge 3 - X, Y2 = 2*X \wedge 2 + 4, Y3 = 2*X - 8$. On the home screen, enter "$\{Y1, Y2, Y3\}$" and store the expression to LF. Of course, we could have avoided the Y= menu by entering the expressions directly into the list on the home screen that defines LF. The two methods are basically equivalent. In the first case, however, we can access the $\vec{j}$ component of $\vec{f}$ by entering Y2 on the home screen, while in the second case, we must enter LF(2) to gain access to the $\vec{j}$ component of $\vec{f}$. (★86★ Press [2ND] [LIST] [F5] [MORE] [MORE] [MORE] [F4] to paste the $Form(\ )$ command onto the active line. Enter

$$Form("\{y1, y2, y3\}", F)$$

and press [ENTER]. The double quote is in the STRING menu. ★86★) There are other ways to enter a vector valued function on our calculator. One of them may seem convenient, but it is not. Surprisingly, we can enter lists as values for the functions in the Y= menu. It is not recommended, but we could have entered $Y1 = \{5*X \wedge 3 - X, 2*X \wedge 2 + 4, 2*X - 8\}$. This behaves like a vector valued function, **but it is not possible to separate the components of $Y1$**, and access to the individual components of a vector valued function turns out to be very important as we continue.

The "listable" property of various operations over lists has helped us enormously in our work with vectors, but our good fortune is about to expire as we begin our work with differentiation and integration. One would hope that the expression $nDeriv(\text{\tiny L}F, X, X)$, for a vector valued function $\text{\tiny L}F$ like the one defined above, would compute the corresponding list of derivatives, but alas, this is not the case. (★86★ Curiously enough, the command $nDer(\ )$ appears to be listable in this sense, but the commands $der1(\ ), der2(\ ), fnInt(\ )$ are not. If you can remember this, it can sometimes be convenient to use this property, but it is an isolated attribute of only one of the calculus commands, we will not attempt to use it . ★86★) To compute the (vector valued) derivative of $\vec{f}(x)$, we must use the $nDeriv(\ )$ command on each component of $\vec{f}(x)$. This could be a very tedious operation, but using the $seq(\ )$ command reduces the keyboard activity substantially. We enter

$$"seq(nDeriv(\text{\tiny L}F(J), X, X), J, 1, 3)",$$

and store the result as $\text{\tiny L}DF$ ("D" for Derivative of F), which gives us the derivative $\vec{f}'(x)$ of $\vec{f}$ (as a function). Again, the quotation marks are critical. For example, press $\boxed{3}$ $\boxed{\text{STO}\blacktriangleright}$ $\boxed{\text{X,T,}\theta,n}$ to set $X = 3$, and then enter $\text{\tiny L}DF$ to compute

$$\vec{f}'(3) = \text{\tiny L}DF = \{134.000005, 12, 2\}.$$

The actual value of $\vec{f}'(3)$, is easily seen to be $\vec{f}'(3) = \{134, 12, 2\}$. It is important to remember, that the $nDeriv(\ )$, command will always compute a value for the derivative, **even if the derivative does not exist.**

In the same way, to compute the (vector valued) antiderivative or integral of a continuous vector valued function $\vec{f}$ like the one defined above, we enter

$$"seq(fnInt(\text{\tiny L}F(J), X, 0, X), J, 1, 3)",$$

and store the result as $\text{\tiny L}AF$ ("A" for Antiderivative of F). The lower limit of integration was set equal to 0, but for this function, which is continuous everywhere, any other value could have been used as well. Recall that if $g(x)$ is continuous on an interval I, and if a is any constant in I then

$$G(x) = \int_a^x g(t)\, dt$$

represents an antiderivative of $g(x)$. Using the function $\vec{f}(x)$ defined above with $X = 3$, enter $\text{\tiny L}AF$ to get

$$\int_0^3 \vec{f}(x)\, dx = \left\{ \int_0^3 (5x^3 - x)\, dx, \int_0^3 (2x^2 + 4)\, dx, \int_0^3 (2x - 8)\, dx \right\} = \{96.75, 30, -15\}.$$

It is easily seen that this value is an exact determination.

Notice that we used $\text{\tiny L}F(J)$ ($J = 1, 2, 3$) in our formulas for $\text{\tiny L}DF$ and $\text{\tiny L}AF$ to access the three components of $\text{\tiny L}F$. While $Y1, Y2, Y3$ are legitimate names for the components of $\text{\tiny L}F$, they cannot be indexed inside of the $seq(\ )$ command, and so they could not be used in this way.

10.1. BASIC CALCULATIONS

Example 10.1 *Let Γ be the curve defined by the position function*

$$\vec{r}(t) = (e^{\sqrt{t}})\vec{i} + \left(\frac{t^2}{t^2+4}\right)\vec{j} + (13\sqrt{t^3+8})\vec{k}.$$

Find the parametric equation of the line tangent to the curve at the point on the curve corresponding to $t = 5.3$.

The point of tangency is $\vec{r}(5.3)$, (as a point, of course) and the direction vector for the line is $\vec{r}\,'(5.3)$. It is convenient to leave our calculator in FUNCTION mode and to use X instead of T. We enter $Y1 = e \wedge (\sqrt{(X)})$, $Y2 = X \wedge 2/(X \wedge 2 + 4)$, $Y3 = 13 * \sqrt{(X \wedge 3 + 8)}$. **Here is an interesting point, which can make our keyboard activity more convenient. It might be more "symbolically" appropriate to enter "$\{Y1, Y2, Y3\}$" on the home screen and store it to $_LR$, but we will intentionally not do this.** From our previous work, $_LF$ is already defined as "$\{Y1, Y2, Y3\}$", and $_LDF$ is already defined as its derivative. We changed our formulas for $Y1$, $Y2$, and $Y3$, but because of our use of the double quote symbols, the formulas for $_LF$ and $_LDF$ are not only updated for the current value of X, but they are updated for $Y1$, $Y2$, and $Y3$ as well. Consequently, the formulas for $\vec{r}(t)$ and $\vec{r}\,'(t)$ are entered as soon as we change $Y1$, $Y2$, and $Y3$.

Set the value of X to $X = 5.3$, enter $_LF$ and store the value to $_LP$ to get the point

$$P = \vec{r}(5.3) = {_LF} = \{9.995878786, .8753505765, 162.8257136\}.$$

Enter $_LDF$ and store its value to $_LV$ to get

$$\vec{v} = \vec{r}\,'(5.3) = {_LV} = \{2.170966148, .0411743207, 43.73274236\}.$$

The vector equation of the line is $\vec{PX} = t\vec{v}$, or $\vec{X} = \vec{P} + t\vec{v}$, where $X(x, y, z)$ denotes an arbitrary point on the line. In parametric form this becomes

$$x = 9.995878786 + 2.170966148\,t$$

$$y = 0.8753505765 + .0411743207\,t$$

$$z = 162.8257136 + 43.73274236\,t.$$

If there were calculations to perform with this line we would probably want to store it on our calculator. Enter "$_LP + X * {_LV}$" and store the result to $_LTN$ to get a "formula" for the tangent line. ∎

Example 10.2 *Create a table of values for the points on the curve Γ defined by the vector valued position function*

$$\vec{r(t)} = \sin(2t)\vec{i} + (\sin(t) - t + 2\cos(3t))\vec{j} + (2t - 3\cos(t))\vec{k}, \quad -2\pi \le t \le 2\pi.$$

Center the table values at $t = 3$ in increments of 0.2 units. Find the equation of the plane which passes through the point $\vec{r(3)}$ on Γ and which is parallel to both $\vec{r}\,'(3)$ and $\vec{r}\,''(3)$.

Enter $Y1 = \sin(2*X)$, $Y2 = \sin(X) - X + 2*\cos(3*X)$, $Y3 = 2*X - 3*\cos(X)$. A table of values for points on the curve may well be of limited use, but it is offered since we have no means to graph the curve. Actually, in one form, a table of values is available without any special preparation. Press [2ND] [TBLSET] set TblStart= 3 and ΔTbl= 0.2, and press [2ND] [TABLE] to see the ordered triplets of point on the curve. While this table is accessible without effort, it does have a drawback. Only two of the three values that define a point in space are visible at the same time. The first entry is the parametric value X (which plays the role of t). To see the third value, one has to scroll to the right or left.

There is a way to create a better looking table—one which displays all three coordinated of the points in the table (at the same time). Enter

$$seq(Y1(1+0.2*J), J, 1, 20),$$

and store the result to, let's say XX (to remind us that it is a list of the x-coordinates of points on the curve). Now press [2ND] [ENTRY], **which brings the last input statement back to the active line where it can be quickly and easily edited** to produce

$$seq(Y2(1+0.2*J), J, 1, 20).$$

Store this value to YY, then in a similar way, enter

$$seq(Y3(1+0.2*J), J, 1, 20),$$

and store it to ZZ. As you can see from the last three arguments in these sequence statements, the index J goes from 1 to 20. The index does not have to start at 1, but there is a certain advantage to having it start there. If $1 \leq k \leq 20$ is an integer, then L$XX(k)$ is the kth term in LXX (regardless of where the index value J starts). As long as our index also starts at 1, we are assured that L$XX(k)$ is the term in the sequence corresponding to $J = k$.

We will place our table on the STAT menu screen. There are several ways to do this, depending on circumstances, but to avoid confusion, we use a procedure that always works. Press [STAT] [5] to paste SetUpEditor onto the home screen. Finish by entering

SetUpEditor XX, YY, ZZ,

and press [ENTER]. This removes old lists from the STAT EDIT screen and places the lists XX, YY, ZZ in that order on the STAT EDIT screen. To view our table of values, press [STAT] [1] (To select Edit...) to see the screen shown consisting of coordinates of points of the curve. The hidden parts of the screen can be viewed by scrolling up or down.

Now, let us find the equation of the plane. If $\vec{r}(t)$ describes the position of an object at time t, then, $\vec{r}'(t)$, and $\vec{r}''(t)$ turn out to be the velocity and acceleration vectors of the object at time t. We will study velocity and acceleration vectors soon enough, but for now, notice, just for the sake of interest, that the plane under

10.1. BASIC CALCULATIONS

consideration seems to describe, in a sense, the plane where all of the "action" is taking place near time $t = 3$, and near the point $\vec{r}(3)$.

To write the equation of the plane, we need a normal vector. Since the vectors $\vec{r}'(3)$ and $\vec{r}''(3)$ are supposed to be in the plane (if they are based at the point $\vec{r}(3)$), it follows that the vector

$$\vec{n} = \vec{r}'(3) \times \vec{r}''(3)$$

is normal to the plane.

If $\mathtt{L}F$ and $\mathtt{L}DF$ are still on our calculator from previous work, they could be used without further preparation to compute $\vec{r}(3)$, and $\vec{r}'(3)$. Actually, we already have $\vec{r}(3)$ on our STAT table, but this vector form is more convenient. For the sake of review, however, let us assume that these formulas have been removed, so that we must start over again. Enter "$\{Y1, Y2, Y3\}$" and store the result to $\mathtt{L}R$. Press $\boxed{3}$ $\boxed{\text{STO}\blacktriangleright}$ $\boxed{\text{X,T,}\theta\text{,}n}$ to set the parameter $X = 3$ (which plays the role of t). Then, the point on the curve corresponding to $t = 3$ is

$$\vec{r}(3) = \mathtt{L}R = \{-.2794154982, -4.681140516, 8.96997749\}.$$

Enter

$$\text{"}seq(nDeriv(\mathtt{L}R(J), X, X), J, 1, 3)\text{"},$$

and store the result to $\mathtt{L}DR$. With $X = 3$, we have

$$\vec{r}'(3) = \mathtt{L}DR = \{1.920339293, -4.462699543, 2.423359954\}.$$

Unfortunately, we cannot compute $\vec{r}''(3)$) in the same way, because we cannot nest the $seq(\ )$ command inside of the $seq(\ )$ command. We will have to apply the $nDeriv(\ )$ command to $\mathtt{L}DR$ by applying it to the three individual terms. Enter

$$nDeriv(\mathtt{L}DR(1), X, 3) \rightarrow A,$$

where A is only for temporary storage. **The remaining two components can then be computed easily** by pressing $\boxed{\text{2ND}}$ $\boxed{\text{ENTRY}}$ and editing the formula to produce

$$nDeriv(\mathtt{L}DR(2), X, 3) \rightarrow B, \ nDeriv(\mathtt{L}DR(3), X, 3) \rightarrow C.$$

We then enter $\{A, B, C\}$ and store the result as $\mathtt{L}D2R$ to get

$$\vec{r}''(3) = \mathtt{L}D2R = \{1.117660503, 16.25917553, -2.969976475\}.$$

Recall that we can only nest the $nDeriv(\ )$ command once inside of the $nDeriv(\ )$ command, and we should probably be somewhat skeptical about how accurately this computes the second derivative. To make these computations more reliable, it would have been better to compute $\vec{r}'(t)$ by hand. We could have used $Y4, Y5, Y6$ to store its values, and they could have been used to define $\mathtt{L}DR$ in much the same way that we defined $\mathtt{L}R$. Let us, however, stay with the easier, although less reliable approach.

We are ready to compute our normal vector $\vec{n}$. Press $\boxed{\text{PRGM}}$, select CROSS, enter $\mathtt{L}DR$ at the first prompt, and $\mathtt{L}D2R$ at the second. Store the resulting value to $\mathtt{L}N$ to get

$$\vec{n} = \mathtt{L}N = \{-26.14772221, 8.411856228, 36.21091664\}.$$

If $P = \vec{r}(3) = {\rm L}R$ represents the fixed point on the plane and $Q = Q(x,y,z)$ represents an arbitrary point on the plane, then the equation of the plane is

$$\vec{n} \cdot \vec{PQ} = 0$$

Expand this as

$$\vec{n} \cdot (\vec{Q} - \vec{P}) = 0,$$

and then write the equation of the plane in the form

$$\vec{n} \cdot \vec{Q} = \vec{n} \cdot \vec{P} = \vec{n} \cdot \vec{r}(3).$$

This makes the right hand side easy to compute by entering $sum({\rm L}N * {\rm L}R)$, and so

$$\vec{n} \cdot \vec{Q} = 292.740105,$$

or in fully expanded form

$$-26.14772221x + 8.411856228y + 36.21091664z = 292.740105.$$

Not all of these digits are significant. Using alternate means, one can establish that each number in the equation of the plane has at least 4 digits of accuracy. ∎

In the next example, we encounter a type of problem which is frequently found in applications.

Example 10.3 *Determine the value $\vec{r}(1.7)$ of a vector valued function $\vec{r}(t)$, such that*

$$\vec{r}''(t) = \cos(3t)\vec{i} + (4 + t^2)\vec{j} - e^{4t}\vec{k},$$
$$\vec{r}'(0) = 2\vec{i} - \vec{j} + 2\vec{k}, \quad \vec{r}(0) = \vec{i} + 6\vec{j} - \vec{k}.$$

Each time we antidifferentiate, we pick up another constant of integration. No big surprise there, but remember that in the current vector setting, the **constant of integration is a vector valued constant**.

We can only use the $fnInt(\ )$ command once, and so we must compute the first antiderivative analytically. A straight forward integration of $\vec{r}''(t)$ gives us

$$\vec{r}'(t) = \vec{g}(t) + \vec{C},$$

where

$$\vec{g}(t) = \frac{1}{3}\sin(3t)\vec{i} + (4t + \frac{1}{3}t^3)\vec{j} - \frac{1}{4}e^{4t}\vec{k},$$

and $\vec{C}$ is a constant vector. From the initial conditions, we have

$$2\vec{i} - \vec{j} + 2\vec{k} = \vec{r}'(0) = \vec{g}(0) + \vec{C},$$

so that

$$\vec{C} = \vec{r}'(0) - \vec{g}(0).$$

Enter $Y1 = \sin(3*X)/3, Y2 = 4*X + X\wedge 3/3, Y3 = e\wedge(4*X)/4$. On the home screen, enter "$\{Y1, Y2, Y3\}$" and store the result to ${\rm L}G$. Enter the initial condition $\{2, -1, 2\}$, store it to ${\rm L}DR0$, enter the initial condition $\{1, 6, -1\}$ and store it to ${\rm L}R0$.

10.2. MOTION IN SPACE

After setting $X = 0$, enter $\text{L}DR0 - \text{L}G$ and store the result to $\text{L}C$. Enter "$\text{L}G + \text{L}C$" and store the result to $\text{L}DR$, which gives us the function $\vec{r}'(t) = \text{L}DR$. Enter

$$"seq(fnInt(\text{L}DR(J), X, 0, X), J, 1, 3)",$$

and store the value to $\text{L}H$. This represents a vector valued function $\vec{h}(t)$ such that

$$\vec{r}(t) = \vec{h}(t) + \vec{K},$$

for some constant vector $\vec{K}$. At the initial point $t = 0$ (that is to say, $X = 0$), clearly $\vec{h}(0) = \vec{0}$, and so $\vec{K} = \vec{r}(0) = \text{L}R0$. With this value for $\vec{K}$, we now have our vector valued function $\vec{r}(t)$. To store this on our calculator, enter "$\text{L}H + \text{L}R0$" and store the result to $\text{L}R$. We compute the value of $\vec{r}(1.7)$ by setting $X = 1.7$, and by entering $\text{L}R$ to get

$$\vec{r}(1.7) = \text{L}R = \{4.469113584, 10.77600833, 58.02795573\}. \quad \blacksquare$$

10.2 Motion in Space

If $\vec{r}(t)$ is the position of an object at time t, then $\vec{v}(t) = \vec{r}'(t)$ is the object's velocity, and $\vec{a}(t) = \vec{v}'(t)$ is its acceleration at time t. The speed of the object at time t is $s(t) = |\vec{v}(t)|$. Our calculator can be an excellent tool to use in the solution of motion problems. We begin with a fairly straightforward example, and follow that with a more interesting one.

Example 10.4 *A golf ball is hit with an initial speed of 140 feet per second, and with an angle of inclination of 34°. The ball travels down a field which is level for the first 374 feet, and which then rises at an angle of inclination of 22°. How far up the hill does the ball land, and what is its speed at impact?*

This can be realized as a two dimensional problem with a vertical y-axis. Let $\vec{g} = -32\vec{j}$ denote the gravitational vector, and $\vec{v}_0 = 140\cos(34°)\vec{i} + 140\sin(34°)\vec{j}$ denote the initial velocity vector. The velocity vector $\vec{v}$ and the position vector $\vec{r}$ of the ball at time t are described by

$$\vec{v} = \vec{v}(t) = t\vec{g} + \vec{v}_0, \qquad \vec{r} = \vec{r}(t) = \frac{t^2}{2}\vec{g} + t\vec{v}_0.$$

Both of these formulas follow by simple antidifferentiation processes, because the acceleration vector $\vec{g}$ is a constant. The hill, which is 374 feet down the fairway, is described by the vector valued function

$$\vec{h} = \vec{h}(s) = 374\vec{i} + s(\cos(22°)\vec{i} + \sin(22°)\vec{j},$$

where the parameter s represents the distance up the hill. At the moment of impact,

$$\vec{r}(t) = \vec{h}(s),$$

and equating the $\vec{i}$ and $\vec{j}$ components of this vector equation gives us two equations in two unknowns, which we can then solve for s and t to give us the time t of impact, and the distance s up the hill where the ball lands.

Equating the $\vec{i}$ components gives us an easy solution

$$s = \frac{140t\cos(34°) - 374}{\cos(22°)}$$

for s in terms of t. This can be substituted into the equation involving the $\vec{j}$ components to give us an equation in just the one variable t.

It turns out that the EQUATION SOLVER does not work well with vector valued functions, so we will use the Y= menu in parametric mode to model these vectors.

We begin, of course, by pressing [MODE] and selecting Par on the fourth line. The [X,T,θ,n] will now produce a T instead of X. To avoid confusion, we will use X_{1T} to represent the solution for s in terms of t that we determined above. In the Y= menu, we enter $X_{1T} = (140 * T * \cos(34°) - 374)/\cos(22°)$ to represent s. Press [2ND] [ANGLE] [1] to paste the degree symbol at the cursor location. (★86★ Press [2ND] [MATH] [F3] [F1]. ★86★) To describe the path $\vec{r}$ of the ball, enter $X_{2T} = 140 * \cos(34°)$, $Y_{2T} = -16 * T \wedge 2 + 140 * \sin(34°)$. To describe the hill 374 feet down the fairway, enter $X_{3T} = 374 + X_{1T} * \cos(22°)$, $Y_{3T} = X_{1T} * \sin(22°)$, where $s = X_{1T}$ expresses s in terms of t.

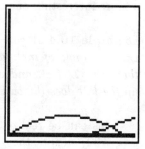

The path of the ball, and the hill down the fairway can now be graphed. Press [WINDOW] and enter $T_{min} = 0$, and $T_{max} = 8$, which is a guess, but can you imagine a golf ball being in the air for more than 8 seconds? Enter $X_{min} = 0$, and (thinking of your favorite golfer) $X_{max} = 900$. Finally, set $Y_{min} = 0$ and $Y_{max} = 900$, so that our picture is scaled properly. After viewing the resulting graph, we will see that our golfer is not quite as powerful as we had imagined. The parameters can then be reduced slightly to produce the screen shown.

Press the [TRACE] key and move the trace cursor along the path of the ball until it meets the hill. This gives us an approximate impact time of $t \approx 4.3$ and an approximate impact location of $(x, y) \approx (488, 46)$. For more accurate readings, press [MATH] [0] to select the SOLVER Screen. We equate the second components of $\vec{r} = \vec{h}$ by entering $eqn : 0 = Y_{2T} - Y_{3T}$. Press ▼ to move to the $T =$ line, where T is already set to its approximate value. Press [ALPHA] [(SOLVE)] to get the impact time of $t = T = 4.2069718011557$ sec.

Back on the home screen, we can enter $\{X_{2T}, Y_{2T}\}$ and press [ENTER] to get the coordinates $(x, y) = (488.2832766, 46.17344091)$ of the impact point in units of feet. Recall that these parametric functions of t are evaluated by pressing [VARS] ▶ [2] and then selecting the desired function. Of course, we could have entered $\{X_{3T}, Y_{3T}\}$ just as well. The "walking distance" up the hill is $s = X_{1T} = 123.2584843$ $feet$.

10.2. MOTION IN SPACE

The impact speed is the length of the velocity vector at this moment in time. Leaving T at the impact time, enter $\vec{v} = t\vec{g} + \vec{v_0}$ in the form

$$\{140 * \cos(34°), 140 * \sin(34°) - 32 * T\},$$

and store the result to $\text{\tiny L}V$. Enter

$$\sqrt{(sum(\text{\tiny L}V \wedge 2))}$$

and press ENTER to get an impact speed of 129.0151145 feet per second. ∎

Example 10.5 *An enemy space ship is spotted traveling along a wide sweeping arc. Captain Ralph hits the attack button on his console and his computer determines the enemy ship's current path, relative to a certain coordinate system expressed in kilometers. The curve appears on his computer screen defined parametrically as*

$$x = 3.1t + 85\cos(.01t) + 400, \quad y = -12.8t, \quad z = 13\cos(.03t) - 4.1t,$$

where t is time in seconds since he hit the attack button. The computer also determines the current path of his own fighter ship and displays the curve on his computer screen parametrically as

$$x = 340 - 2.1t, \quad y = 197\sqrt{t+1} - .015t^2 - 2300, \quad z = 0.032t^2 + 3.6t - 608.$$

"What a great computer," Ralph says to himself, as he prepares to launch a missile and attack the enemy ship.

First he sets the missile launch time to $t = 40$ seconds, and then he tries to determine the missile launch direction at $t = 40$. The missile will launch with a speed of 20.8 kilometers per second (relative to Ralph's ship). The missile has no power of its own, and so it will travel in a straight line with a constant velocity vector until it bumps (hopefully) into the enemy ship and explodes. What direction vector should Captain Ralph give the missile so that it will knock out the enemy ship? At what time will the missile reach the enemy ship, and how far will it be from Captain Ralph at that moment? Ralph, unfortunately, is a "shoot from the hips kind of guy," and not a very good student of calculus. He really needs our help.

We let $\vec{e}(t)$, $\vec{r}(t)$, and $\vec{m}(t)$, denote the vector valued position functions for the <u>e</u>nemy ship, <u>R</u>alph's ship, and the <u>m</u>issile respectively. These letters will be prefixed with v to denote respective velocities. If Ralph points the missile in the direction of some unit vector $\vec{u}$, then the velocity vector of the missile will be the constant vector

$$v\vec{m} = 20.8\vec{u} + v\vec{r}(40).$$

The reasoning here is that the missile will acquire not only the launch velocity relative to Ralph, but the velocity of Ralph's ship at the firing time as well. When we need it, we can easily compute $v\vec{r}(40)$ using our calculator and the ideas of calculus, so for the time being, let us just treat it as a known quantity. We need to determine the vector $\vec{u}$ (three unknowns), and the time t (a fourth unknown) when $\vec{m}(t) = \vec{e}(t)$ (when the missile and the enemy ship are at the same position).

With a constant velocity vector, the position vector $\vec{m}(t)$ of the missile is a simple antiderivative of $v\vec{m}$. We know that at time $t = 40$, the position of the missile should be $\vec{r}(40)$ (Ralph's position at that time). This is a vector we can easily compute when we need it. We could write the vector $\vec{m}(t)$ in the form

$$\vec{m}(t) = t\, v\vec{m} + \vec{C},$$

where the constant $\vec{C}$ of integration could be evaluated from the initial condition. However, it is more convenient to use $t - 40$ instead of t and write

$$\vec{m}(t) = (t - 40)v\vec{m} + \vec{r}(40).$$

Notice that this expression for $\vec{m}(t)$ is clearly an antiderivative of $v\vec{m}$ and it clearly satisfies $\vec{m}(40) = \vec{r}(40)$. This is all we need to conclude that it must be the correct expression for $\vec{m}(t)$.

The missile will bump into the enemy ship when $\vec{m}(t) = \vec{e}(t)$. This gives us

$$\vec{m}(t) = (t - 40)v\vec{m} + \vec{r}(40) = \vec{e}(t).$$

Using our expression for $v\vec{m}$, we write

$$\vec{m}(t) = (t - 40)(20.8\vec{u} + v\vec{r}(40)) + \vec{r}(40) = \vec{e}(t).$$

We solve this equation for the unknown vector $\vec{u}$ to get

$$\vec{u} = \frac{1}{20.8(t - 40)}(\vec{e}(t) - \vec{r}(40)) - \frac{1}{20.8}v\vec{r}(40).$$

From this vector equation, we can express the components u_1, u_2, u_3 of $\vec{u} = u_1\vec{i} + u_2\vec{j} + u_3\vec{k}$ in terms of t. Since $\vec{u}$ is intended to be a unit (direction) vector, these expressions in t can then be substituted into the unit length equation

$$u_1^2 + u_2^2 + u_3^2 = 1$$

to get one equation in the one unknown t, which can subsequently be solved for t.

Our calculator work begins by entering the vector valued function $\vec{r}$, and evaluating $\vec{r}(40)$ and $v\vec{r}(40)$. We put our calculator in function (Func) mode and let X represent the time variable t. On the home screen, enter Ralph's position function

"$\{340 - 2.1 * X, 197\sqrt{(X + 1)} - .015 * X \wedge 2 - 2300, .032 * X \wedge 2 + 3.6 * X - 608\}$",

and store the result to LR. Set $X = 40$ ($X = t$) and enter LR to get the position of Ralph and the missile at launch time as

$$\vec{r}(40) = \text{L}R = \{256, -1062.584525, -412.8\}.$$

This vector is stored as L$R40$. (We need this fixed vector, and LR will change as X changes.) To compute the velocity vector of Ralph's ship at time $t = 40$ we enter

$$seq(nDeriv(\text{L}R(J), X, 40), J, 1, 3)$$

10.2. MOTION IN SPACE

and store the result to $_LVR$ to get

$$\vec{vr}(40) = {}_LVR = \{-2.1, 14.1831155, 6.16\}.$$

We use these values, along with

$$\vec{e}(t) = (3.1t + 85\cos(.01t) + 400)\vec{i} + (-12.8t)\vec{j} + (13\cos(.03t) - 4.1t)\vec{k}$$

in the vector equation for $\vec{u}$ above to get our expressions for u_1, u_2, u_3. Notice that if $\vec{e} = \vec{e}(t)$, $\vec{r} = \vec{r}(40)$, $\vec{vr} = \vec{vr}(40)$ are simplified notationally to just

$$\vec{e} = e_1\vec{i} + e_2\vec{j} + e_3\vec{k}, \quad \vec{r} = r_1\vec{i} + r_2\vec{j} + r_3\vec{k}, \quad \vec{vr} = vr_1\vec{i} + vr_2\vec{j} + vr_3\vec{k},$$

then the expressions for u_1, u_2, u_3 are all simply

$$u_j = \frac{1}{20.8(t-40)}(e_j - r_j) - \frac{1}{20.8}vr_j \quad (j = 1, 2, 3).$$

This makes it easy to write these expressions by observation as

$$u_1 = \frac{1}{20.8(t-40)}(3.1t + 85\cos(.01t) + 400 - 256) + \frac{2.1}{20.8},$$

$$u_2 = \frac{1}{20.8(t-40)}(-12.8t + 1062.584525) - \frac{14.1831155}{20.8},$$

$$u_3 = \frac{1}{20.8(t-40)}(13\cos(.03t) - 4.1t + 412.8) - \frac{6.16}{20.8}.$$

We enter these formulas for u_1, u_2, u_3 on our calculator as $Y1, Y2, Y3$. (**Remember that we are using X to represent t**). This tedious (sorry) but obvious step is omitted from our presentation.

Our immediate objective is to solve the unit length equation $Y1 \wedge 2 + Y2 \wedge 2 + Y3 \wedge 2 = 1$ for X ($X = t$). We can enter this equation into our EQUATION SOLVER, but what X-value should we use for a start up solution? How long approximately does it take for the missile to (hopefully) bump into the enemy ship? We have no knowledge of this, but we can answer this question with a graph. We enter $Y4 = Y1 \wedge 2 + Y2 \wedge 2 + Y3 \wedge 2 - 1$. Press [WINDOW], set $X_{min} = 0$, $X_{max} = 100$ (a guess), and since we are only interested in where $Y4 = 0$, set $Y_{min} = -2$, and $Y_{max} = 2$, to produce the screen shown. Press [TRACE] and move the trace cursor to where the graph crosses the X-axis, to get an approximate value of $X \approx 61$.

With this approximate solution, we press [MATH][0] to bring up the SOLVER screen. Enter $eqn : 0 = Y4$, press ▼ (X will already have the X-value of the trace cursor, and press [ALPHA] ([SOLVE]) to get $t = X = 61.74999723478$. This is the impact time in seconds, when the missile and the enemy ship meet. The direction vector is

$$\vec{u} = \{u_1, u_2, u_3\} = \{Y1, Y2, Y3\} = \{.99558556912, -0.0802346753, 0.0486983395\},$$

which we now store to ʟU. The position of the missile at impact can be found using the formula

$$\vec{m}(t) = (t-40)v\vec{m} + \vec{r}(40) = (t-40)(20.8\vec{u} + v\vec{r}(40)) + \vec{r}(40)$$

obtained earlier, with $t = X$ fixed at our impact time (it already has the correct value). We enter

$$(X-40) * (20.8 * \text{ʟ}U + \text{ʟ}VR) + \text{ʟ}R40,$$

and store the value to ʟP to get an explosion point (in units of kilometers) of

$$\text{ʟ}P = \{660.7279153, -790.3999637, -256.7888911\}.$$

At that moment Ralph's location (in units of kilometers) can be found by leaving X unchanged and entering ʟR on the home screen to get

$$\text{ʟ}R = \{-210.3250058, -796.6624843, -263.682020\}.$$

The distance between Ralph and the explosion point is the length of the vector $\vec{r} - \vec{P}$ (at impact time, of course). Recall that the length of a vector $\vec{w}$ is computed on our calculator by using the rule $|\vec{w}| = \sqrt{\vec{w} \cdot \vec{w}}$. Without changing X, we enter

$$\sqrt{(sum((\text{ʟ}R - \text{ʟ}P) \wedge 2))}$$

to get a distance between Ralph and the explosion point of $|\vec{r} - \vec{P}| = 450.4991845$ kilometers. Nice shooting Captain Ralph! ∎

10.3 Geometry of Curves

The ideas of slope and concavity that we used to describe the geometry of the graph of a real valued function of a real variable are ideas that can be used to describe the geometry of a parametrically defined curve in two or three dimensional space in much the same way. The geometry is described by the so-called "curvature" of a curve, and by three vectors: the unit tangent vector $\vec{T}$, the principal unit normal vector $\vec{N}$, and the binormal vector $\vec{N}$.

Let Γ be a curve defined by the continuous vector valued function

$$\vec{r} = \vec{r}(t), \ a \leq t \leq b. \tag{10.1}$$

In order to study just the pure geometric properties of the curve, without dealing with the motion of the point $\vec{r}(t)$ moving along the curve, it is useful to strip away the idea of the speed at which the point $\vec{r}(t)$ moves along the curve. For this reason, we use **the unit tangent vector**

$$\vec{T} = \vec{T}(t) = \frac{1}{|\vec{r}'(t)|}\vec{r}'(t) = \frac{1}{|\vec{v}(t)|}\vec{v}(t), \tag{10.2}$$

rather than the velocity vector $\vec{v}(t) = \vec{r}'(t)$ to describe the geometric properties of tangency. Think of $\vec{T}(t)$ as the velocity vector of a point which is moving along the curve with a constant unit speed.

10.3. GEOMETRY OF CURVES

If a object, traveling along Γ, simply slows down and stops, quite smoothly, at some point on the curve, the unit tangent vector can not be computed from $\vec{r} = \vec{r}(t)$. The curve could be quite smooth at the point with a well defined tangent line, or it could have a sharp corner at the point. That is to say, a continuous velocity (tangent) vector $\vec{r}\,'(t)$ can move around a sharp corner at a point $\vec{r}(t_0)$ on a curve, as long as $\vec{r}\,'(t_0) = \vec{0}$. Continuity of $\vec{r}\,'(t)$ does not rule out sharp corners. Think of a car moving down a road. It can negotiate a sharp corner in the road as long as it first stops.

This annoyance is best resolved by starting with a parametrization for which $\vec{r}\,'(t)$ is never zero, except possible at the end points. We say that the curve Γ defined by 10.1 is smooth, if

$$\vec{v} = \vec{v}(t) = \vec{r}\,'(t),$$

is continuous and nonzero on the interval $a \leq t \leq b$ (except possibly at the endpoints.) **A smooth curve cannot have sharp corners.**

The **principal unit normal vector** (or just unit normal vector) to the curve Γ defined by 10.1 is

$$\vec{N} = \vec{N}(t) = \frac{1}{|\vec{T}\,'(t)|}\vec{T}\,'(t) = \frac{\vec{T}\,'(t)}{|\vec{T}\,'(t)|}.$$

The last of the three vectors, the **binormal vector** is defined by

$$\vec{B} = \vec{T} \times \vec{N}.$$

For each t, the vectors $\vec{T}$, $\vec{B}$, and $\vec{N}$ are mutually orthogonal unit vectors (a so-called orthonormal set). Why? Actually, the answer is immediate—no computations are required to verify this result. **The collection $\{\vec{T}, \vec{B}, \vec{N}\}$, of vectors based at the point $\vec{r}(t)$ on the curve Γ is called the frame (or TNB frame) based at $\vec{r}(t)$.** This is a frame of mutually orthogonal unit vectors that moves along the curve with the point.

To study the geometry of a curve, it is useful to strip away, even further, the vestiges of time from its parameter. Your calculus text will discuss the notion of reparametrizing a curve in terms of the arc length variable s. Simply put, if Γ has arc length L, and $0 \leq s \leq L$, then $\vec{r} = \vec{r}(s)$ is parametrized in terms of arc length, if the point $\vec{r}(s)$ on the curve corresponding to s, is the point which is exactly s units along the curve from the initial point $\vec{r}(a)$. Your calculus text will define this more carefully, and show the relationship between the original parameter, and the arc length parameter. We should not, however, let the technical details get in the way—the idea, itself, is intrinsically very simple and natural. Simply put, if the curve $\vec{r} = \vec{r}(s)$ were a highway, paramatrized in terms of arc length s in miles, then $\vec{r}(s)$ is the point on the highway s miles down the road.

Arc length as a parameter is a very important concept, but, as it turns out, it is generally very hard, given a curve, to actually find a formula which paramatrizes a curve in terms of arc length. Fortunately, it is the idea itself, and not the formula, which makes this an important concept, and ideas in terms of arc length can always be turned into formulas in terms of the given parametric variable t. Your calculus text will probably show that if the curve Γ defined by 10.1 is reparametrized in terms of arc length s, then the unit tangent vector can be expressed as

$$\vec{T} = \frac{d\vec{r}}{ds}. \qquad (10.3)$$

This is a very satisfying result. It means, for example, that if we think of the arc length parameter as a new "time-like" variable, then motion along the curve always has unit speed with this new parameter. To compute $\vec{T}$, on the other hand, we will still have to use 10.2, since finding a formula for the curve in terms of arc length is usually beyond our reach. Arc length as a parameter is used to build ideas, not to compute with.

The **curvature** of a curve is a measure of how quickly the unit tangent vector is turning. We do not want this to depend on the speed of a point moving along a curve, and so curvature is defined by

$$\kappa = \left|\frac{dT}{ds}\right| = \left|\frac{dT}{dt}\frac{dt}{ds}\right| = \frac{1}{|\vec{r}\,'(t)|}\left|\frac{dT}{dt}\right|.$$

Notice that $\frac{ds}{dt} = \vec{r}\,'(t)$ is just the speed of the point $\vec{r}(t)$ moving along the curve. Finally we remark that the formula

$$\vec{N} = \frac{1}{\kappa}\frac{dT}{ds} = \frac{1}{\kappa}\frac{dT}{dt}\frac{dt}{ds} = \frac{1}{\kappa}\frac{dT}{dt}\frac{1}{|\vec{r}\,'(t)|}$$

is sometimes useful.

Before we start our next example, let us take a moment to delete the large number of lists (vectors) that are probably accumulating in our calculator's memory. Press [2ND] [MEM] [2] [4] and delete each unwanted list by moving the cursor to the list and pressing [ENTER].

Example 10.6 *Let Γ be the curve defined by the vector valued position function*

$$\vec{r} = \sin(t)\vec{i} + (\cos(t) - \sin(t))\vec{j} + \cos(3t)\vec{k},\ 0 \leq t \leq 2\pi.$$

Compute the vectors $\vec{T}$, $\vec{N}$, $\vec{B}$ at the point on Γ corresponding to $t = \pi/4$. Determine the curvature of Γ at $t = \pi/4$.

Some attention has to be focused on the evaluation procedure. It is clear from the above formulas that we must compute $\vec{T}$ as a vector valued function of t, so that we can differentiate $\vec{T}$ to get $\vec{N}$. This is where computations frequently get out of hand. Once we differentiate $\vec{T}$, there is no longer a need to maintain t, as a variable, and so we immediately evaluate everything at $t = \pi/4$ to avoid complications.

Let us place our calculator in function (Func) mode, and let X represent t. Enter $\pi/4 \to X$ and press [ENTER]. The value of X will remain here for the rest of the problem, although X must also play the role of a variable. This double meaning for X can be confusing. Recall that an entry of the form "$E(X)$" (one enclosed inside of double quotes) which is stored to a letter F essentially turns F into a function of the variable X.

10.3. GEOMETRY OF CURVES

We must differentiate twice to get $\vec{N}$, and a double application of the $nDeriv(\ )$ command can be avoided by computing the first derivative $\vec{r}'(t)$ by hand. This is surely the only reasonable approach. The first derivative is easy to compute, and our final calculations will be more accurate, or at least more reliable.

Even without a calculator, it is easy to compute the exact coordinates of the point

$$\vec{r}(\pi/4) = \left\{\frac{1}{\sqrt{2}}, 0, -\frac{1}{\sqrt{2}}\right\} = \{.7071067812, 0, -.7071067812\}$$

on the curve Γ corresponding to $t = \pi/4$. There is no need to enter this vector valued function on our calculator. Computing $\vec{r}'(t)$ by hand, we enter

$$"\{\cos(X), -\sin(X) - \cos(X), 3*\sin(3*X)\}",$$

and store the result as $\text{L}V$. Because of the double quotes, it follows that $\text{L}V = \vec{v} = \vec{r}'$ is a vector valued function of X ($X = t$). Using the definition of $\vec{T}$ given in the introduction, we form $\vec{T}$ by turning $\vec{v} = \text{L}V$ into a unit vector. Enter

$$"\text{L}V/\sqrt{(sum(\text{L}V \wedge 2))}",$$

and store the result to $\text{L}T$. Notice that $\text{L}T$ is also entered as a vector valued function—a critical step, since we must differentiate this function. Enter $\text{L}T$ and press ENTER (X is already $\pi/4$) to get

$$\text{L}T = \vec{T} = \{.2672612419, -.5345224838, -.801783725\}.$$

From this point forward, all of our vectors will be evaluated at $X = \pi/4$, so we will no longer need the double quote symbols in our work. Enter

$$seq(nDeriv(\text{L}T(J), X, X), J, 1, 3),$$

and store the result to $\text{L}DT$ (for the Derivative of $\vec{T}$) to get

$$\text{L}DT = \vec{T}' = \{.2672618464, -1.069049371, .8017947026\}.$$

The first X in the above $nDeriv(\ )$ command is the name of the variable used in the differentiation process. The second X (since we're not using double quotes) is just the current value of X, namely $X = \pi/4$. From the definition given in the introduction, we compute the principal unit normal vector by turning $\text{L}DT$ into a unit vector. Enter

$$\text{L}DT/\sqrt{(sum(\text{L}DT \wedge 2))},$$

and store the result to $\text{L}N$ to get

$$\text{L}N = \vec{N} = \{.1961151352, -.7844619975, .5883521295\}.$$

Finally, to get the binormal vector $\vec{B} = \vec{T} \times \vec{N}$, we press PRGM and choose CROSS in the EXEC menu. At the first prompt, enter $\text{L}T$ and at the second prompt, enter $\text{L}N$. Store the result to $\text{L}B$ to get

$$\text{L}B = \vec{B} = \vec{T} \times \vec{N} = \{-.9434563047, -.3144856445, -.1048283385\}.$$

The vector $\vec{B} = {}_\text{L}B$ automatically has unit length. Why? This completes the computation of the TBN frame based at the point $\vec{r}(\pi/4)$.

A definition of curvature was given in the introduction. Using the formula in this definition, we enter

$$\sqrt{(sum({}_\text{L}DT \wedge 2))}/\sqrt{(sum({}_\text{L}V \wedge 2))},$$

to get a curvature of $\kappa = .515082545$ at the point $\vec{r}(\pi/4)$ on the curve Γ. ∎

If this curve were plotted with a 3-dimensional plotting device, it would be observed that in addition to being perpendicular to Γ, $\vec{N}$ **always points to the concave side of the curve**, towards the center of what we call the circle of curvature.

The radius of curvature is $\rho = 1/\kappa$. If Γ is the curve defined by 10.1, and if $\vec{T}$, $\vec{N}$, $\vec{B}$, κ, and ρ are all evaluated at the point $\vec{r}_0 = \vec{r}(t)$ on Γ, then the circle of curvature is the circle with radius ρ, centered at the point $\vec{r}_0 + \rho\vec{N}$, which lies in the plane through $\vec{r}_0$ with normal vector $\vec{B}$ (the plane formed by $\vec{T}$, and $\vec{N}$).

If formula 10.1 defines the motion of an object along a curve Γ, then the most useful way to present the acceleration vector, $\vec{a}$, acting on the object at time t is to express it in the form

$$\vec{a} = a_T \vec{T} + a_N \vec{N}.$$

The tangential component of acceleration, a_T, changes the speed of the object, and the normal component of acceleration, a_N, changes the direction of the object. Finally, the object's acceleration is related to the force acting on it by *Newton's Second Law* $\vec{F} = m\vec{a}$. These issues will be raised in the exercise set.

10.4 Exercise Set

1. An object moves along a curve with position vector

$$\vec{r} = (3 - \sin(4t))\vec{i} + 2t\cos(t)\vec{j} + 5\sin(3t)\vec{k}, \ 0 \leq t \leq 10.$$

Find, at time $t = 7$, the velocity, acceleration, and speed of the object, the angle between the velocity and acceleration vectors, and the projection of the acceleration vector onto the velocity vector. Using just this information at time $t = 7$, decide whether the object is speeding up or slowing down at this time.

2. Let γ be a circle in the xy-plane, of radius r centered at the point $A(0, R)$ where $0 < r < R$. A torus is formed by rotating γ about the x-axis. Two circular axes are naturally associated with the torus. One would be γ itself, and the other would be the circle β formed by rotating the point $A(0, R)$ about the x-axis.

If m, and n are relatively prime integers (m, n have no common prime factors other than 1), then the following parametrically defined curve, with domain $0 \leq t \leq 2\pi$ wraps m times around the torus in one "circular" way, and n times around the torus in the other "circular" way.

$$\begin{vmatrix} x = & r\sin(2\pi m t) \\ y = & (R + r\cos(2\pi m t))\cos(2\pi n t) \\ z = & (R + r\cos(2\pi m t))\sin(2\pi n t) \end{vmatrix}$$

10.4. EXERCISE SET

Set $r = 1$, $R = 4$, $n = 3$, $m = 7$ and compute the arc length of the curve. Plan on doing something else for about a half hour after the arc length integral is entered.

3. In a rectangular coordinate system with units in feet, a projectile is shot from a gun at position $P(2, -3, 6)$ into a constant force field with acceleration vector $\vec{a} = 4.2\vec{i} - 1.8\vec{j} + 29.7\vec{k}$ in ft/sec^2. No other forces are acting on the projectile. The muzzle velocity of the gun (the initial speed of the projectile) is $2{,}000 \, ft/sec$, and the gun is pointed in the direction of the vector $\vec{g} = 7.2\vec{i} + 12.4\vec{j} + 8.82\vec{k}$. Find the velocity and position of the particle at time t in seconds. How close does the projectile come to a target located at $Q(120, 213, 89)$? The muzzle velocity of the gun is a constant $2{,}000 \, ft/sec$, but its direction can be altered. Find a direction (expressed as a unit vector) so that the projectile will hit the target.

4. A force (in pounds) of

$$\vec{F} = 8\sqrt{t}\,\vec{i} - 3t\vec{j} - \frac{7}{t^2 + 4}\vec{k}, \quad 0 \le t \le 60$$

is exerted on an object of mass $m = 12 \, slugs$, where t is time in seconds. The position and velocity of the object at time $t = 0$ are

$$\vec{v}_0 = 2\vec{i} + 7\vec{j} + 13\vec{k}, \quad \vec{r}_0 = 8\vec{i} - 33\vec{j} - 9\vec{k}.$$

Find the velocity and position of the object for time t in the interval $[0, 60]$. (Force is related to acceleration by *Newton's Law*, $\vec{F} = m\vec{a}$. When the force $\vec{F}$ is expressed in pounds, and mass m is expressed in slugs, then acceleration $\vec{a}$ is expressed in ft/sec^2.)

5. Let $\vec{f}(t)$, and $\vec{g}(t)$ be vector valued functions. State a *Product Rule* for the derivative of $\vec{f}(x) \cdot \vec{g}(x)$. If this rule is not in your main calculus text, use your background in calculus and your imagination to make a conjecture (a wise guess).
a) Verify that the rule holds at $x = 13$ for

$$\vec{f} = 3xe^x\vec{i} + \arctan(5x)\vec{j} - (8x^3 + 12x^2 + 9)\vec{k},$$
$$\vec{g} = \cos(x^3 + 4)\vec{i} - (x^4 - 4^x)\vec{j} + (x^2 \sin(9x))\vec{k}.$$

b) For a much improved version of part a), set up the solution so that, once it is entered, the rule can be verified, with very minimal keyboard activity, for any chosen value of $x = X$.

6. Let $\vec{f}(x)$, and $\vec{g}(x)$ be vector valued functions. State a *Product Rule* for the derivative of $\vec{f}(x) \times \vec{g}(x)$. If this rule is not in your main calculus text, use your background in calculus and your imagination to make a conjecture (a wise educated guess). Verify that the rule holds at $x = 13$ for the functions $\vec{f}(x)$, and $\vec{g}(x)$ in problem 5.

7. Let $\vec{f}(x)$, be vector valued function, and let $c(t)$ be a scalar (real valued) function. State a *Product Rule* for the derivative of $c(x)\vec{f}(x)$. If this rule is not in your main calculus text, use your background in calculus and your imagination to make a conjecture (a wise educated guess). Verify that the rule holds at $x = 7$ for $c(x) = \left(\ln(x^4 + 18)\right)^3$ and the function $\vec{f}(x)$ in problem 5.

8. A projectile is shot from the top of a 100 ft tall building, at an angle of elevation of 27° (above horizontal) and at an initial speed of $243^{ft}/_{sec}$. How high does the projectile go? What is its speed and angle at impact? How much time does it take to land, and how far down range is it at impact? Assume a level field surrounding the building.

9. A baseball is hit 3 ft above ground, down the third base line, at an angle of elevation of 26°, and at an initial speed of $140^{ft}/_{sec}$. Will it clear a 30 ft. high fence which is 340 ft from home plate down the third base line?

10. Let Γ be the spiral

$$\vec{r} = \vec{r}(t) = a\cos(2\pi nt)\vec{i} + a\sin(2\pi nt)\vec{j} + 2\pi nbt\vec{k}, \ 0 \leq t \leq 1,$$

where a and b are positive constants and n is a positive integer. Show that Γ appears to be a curve of constant curvature κ, and determine the value of κ. Assign values to $a = A$ and $b = B$, and set up κ as a "formula" so that it can easily be evaluated for several values of A, B, and $t = T$. Find the center of the circle of curvature corresponding to the point $\vec{r}(t)$ on Γ for several values of $a = A$, $b = B$, and $t = T$. Before using your calculator to perform these computations, try to guess their expected values. Make a conjecture, and use your computations to verify (partially) your conjecture.

11. Determine the frame vectors $\vec{T}$, $\vec{N}$, and $\vec{B}$ at the point $t = 2/3$ on the curve Γ defined by

$$x = \sin(4\pi t), \ y = (4 + \cos(4\pi t))\cos(10\pi t), \ z = (4 + \cos(4\pi t))\sin(10\pi t)$$

The curve Γ is one of the curves discussed in Problem 2. Determine the curvature, the radius of curvature, and the center of the circle of curvature at the point $t = 2/3$. If t is time, the parametric representation for Γ describes the motion of an object along the curve. Find the tangential, a_T, and normal, a_N components of acceleration at $t = 2/3$.

Project: Slamming Sam's Home Run Attempt

A baseball field is shown in the following figure. The angle at home plate (labeled H in the picture) is 90° and the three outfield fences are all 30 ft. tall. All of the other information concerning the field is shown in the picture. Slamming Sam hits the ball 2.8 ft above ground, at an angle of elevation of 25°, and at an initial speed of 152 ft/sec. From a "bird's eye" view, far above the playing field, the ball travels in a direction exactly midway between second and third base, as shown in the picture. Does Slamming Sam have another home run? If not, describe, in very exact language, how close it is to being a home run.

Project: Captain Ralph's Computer Failure (Version 1)

Captain Ralph is on a routine patrol a few thousand miles away from his home space station where the origin of a three dimensional coordinate system is located. (Coordinates are expressed in miles.) Looking for action, he grins as he

10.4. EXERCISE SET

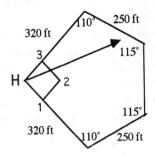

spots an enemy ship. His display monitor shows that his coordinate position is $P(405, -3201, 95)$. He hits the attack button on his computer, and his ship begins to travel, starting at $t = 0$, along the attack curve

$$\vec{r}(t) = (405 - t^2)\vec{i} + (5t - 3201)\vec{j} + (95 + t^3)\vec{k}$$

(t is in seconds), firing off a projectile at the enemy ship $t = 2$ seconds later. The projectile travels along a straight line having the same direction as Captain Ralph's velocity vector at time $t = 2$, with a constant acceleration vector of 1 $mile/sec^2$ (naturally, in the same direction as Captain Ralph's velocity vector at time $t = 2$). The projectile has an initial speed which is 14 miles per second faster than Captain Ralph's speed at the firing time.

Unfortunately, Ralph's computer makes a mistake, and it programs the projectile to explode prematurely 4 seconds later. When the projectile explodes, it will annihilate everything within a 100 mile radius, and in the meantime, Captain Ralph has no choice but to continue along the same attack curve for another 5 seconds, before he regains control of his ship. Will he make it? Will he be 100 miles away from the projectile when it explodes? Come on Captain Ralph! Weeeeeeeeeeee ...neeeeeeeeeed ...youuuuuuuuuu!

Project: Captain Ralph's Computer Failure (Version 2)

This version is the same as the first, except for the way that the bomb explodes. As every fighter pilot for the United Earth Federation knows, bombs do not always explode in spherical patterns. Acceleration warps and flattens the explosive region.

By a circular ellipsoid we mean the surface generated by revolving an ellipse about its major or minor axis. The axis of the circular ellipsoid is the line segment on the axis of rotation terminating at points on the surface. The center of the ellipse is the midpoint of the axis. The diameter of the ellipsoid is the diameter of the circle formed by intersecting the ellipsoid with a plane passing through the center which is perpendicular to the axis. It is a straightforward exercise to show that if d is the diameter, and l is the length of the axis of a circular ellipsoid, then the volume of the enclosed region is

$$V = \frac{\pi l d^2}{6}.$$

As you can see, when $l = d = 2r$ (r =radius), this turns into the familiar formula for the volume of a sphere of radius r.

When Captain Ralph fires a projectile and it travels at a constant velocity, its region of destruction is the region inside a sphere of radius 100 miles centered at the point of explosion. However acceleration warps the region of destruction into a circular ellipsoid centered at the explosion point, having an axis parallel to the acceleration vector. The volume of the region of destruction remains the same, but it flattens so that l decreases, and d increases, according to the rule

$$l = \frac{d}{1 + 2|\vec{a}|},$$

where $\vec{a}$ denotes the acceleration vector expressed in $miles/sec^2$.

Outside of this change in the explosive pattern, this version of the problem is the same as the first. Determine whether or not Captain Ralph survives this computer failure.

You have all the tools needed to solve this problem, but this is definitely a nontrivial project. A planar ellipse is defined in terms of the sum of the distances between a point on the ellipse and its two focal points. This definition can be used very effectively to decide whether a point P is inside or outside an ellipse. Now try to turn this problem into a planar problem.

Epilogue

The remaining topics in multivariate calculus are among the most fasinating topics in all of calculus, and they are certainly among the most applied topics. The world is, afterall, a function of many variables. As you head into the best part of calculus, however, it is time for your calculator to gracefully say good-by (press [2ND] [CATALOG] [G] ..., well, then again, there doesn't appear to be such a command).

To be sure, you will find many occasions to use your calculator in multivariate calculus, and you are encouraged to use it when it is appropriate. The usage will, however, be much more sporadic. There is no technological advantage gained by forcing our calculator to behave unnaturally in a multivariate environment.

If you wish to continue to use technology in a significant way to do mathematical and scientific computation, it is recommended that you select one of the many powerful software applications available for desk top computers.

Finally, it should be said that the calculators we have been using have many features that have not been explored, because this is only a book about calculus. You will find this versatile instrument to be a very useful tool in many of your future mathematical and scientific explorations.

———★ ★ ★———

Index

Indexed items[1]

absolute convergence, 118
acceleration vector, 176, 179
acceleration: normal and tangential components, 188
accuracy: graphical evidence, 4
active function mode, 3
alternating series test, 117, 119
antiderivative: as a function, 52, 80
antiderivative: constant of integration, 52
arc length as parameter, 185
area, 61, 83
area: vertical vs. horizontal approach, 63, 84
asymptotic behavior, 33

binormal vector, 185

calculator limitations, 115
CATALOG menu, 76, 93
center of mass, 66
circle of curvature, 188
comparison tests, 113
conditional convergence, 118
creating unit vectors, 154
creating you own programs, 167
cross product, 157, 158, 162
cross product, nonassociative, 160
curvature, 186
curvature circle, 188
curvature radius, 188
curve smooth, 185

curve-arc length parameter, 185

der1( ) vs. nDer( ) (Ti-86), 24
der2( ) command (Ti-86), 24
derivative
    a warning, 24
    approximation, 22
    as a function, 25
    Chain Rule, 25
    degree of accuracy, 24
    der1( ) command (Ti-86), 24
    formal definition, 22
    graphical definition, 21
    nDer( ) command, 24
    nDeriv( ) command, 24, 25
    nonexistence, 23
    of an inverse function, 79
    Product Rule, 25
    sharp corner, 23
    table of values, 23
    tangent line, 26
    values, 24
derivative vs. integral, 53
differential equations (Ti-86), 92
differentiation
    vector valued functions, 177
distance between a point and a line, 166
distance between a point and a plane, 163
distance between lines, 162
dot product, 157, 158, 162
DRAW menu, 26, 35, 55

e: a definition and value, 81
entering old statements, 38, 39, 50, 58, 67, 124, 141, 155, 167, 176, 177

---

[1] Items are, for the most part, indexed mathematically rather than by calculator commands/keys. See the index in the manual that came with your calculator for a more direct index of calculator commands/keys

entry key, 38, 39, 50, 58, 67, 124, 141, 155, 167, 176, 177
equation of a plane, 162
equation(s)
    comparing solution methods, 38
    complications, 32, 36, 38, 39
    inverse function values, 75
    inverse program, 75
    Newton's Method, 37–39
    polynomial, 32
    SOLVE menu, 7, 27, 32, 36, 55, 62, 64, 69, 81, 82, 89, 137, 180, 183
    solving graphically, 1, 6
    systems of, 165
    vector, 180
$\exp(x)=e^x$: a definition as an inverse, 81
$\exp(x)=e^x$: order of growth, 82
exponential growth, 90
exponential growth and decay, 90
expressions as names, 67

factoring large polynomials, 36
factory setting, 3
Fibonacci sequence, 109
fnInt( ) command, 51, 80
    not listable, 174
    on a vector, 174, 178
    repeated applications, 178
fnInt( ) command, 52
force, 68
FORMAT menu, 135, 140
function(s)
    active, 3, 9
    composition, 12, 25
    decreasing, 25
    inactive, 3, 9, 22, 41
    increasing, 25
    of other functions, 8, 25, 40
    piecewise, 7
    plotting, 3
    values, 7, 8, 12
functions-inaccessible, 101
Fundamental Theorem of Calculus, 51

geometric series, 113
geometry of curves, 184
graphing, *see* plotting
graphing functions, *see* plotting functions
graphing styles, *see* plotting styles
greatest common devisor command, 164

half life, 90
harmonic series, 113, 115
hyperbolic functions, 93

implicit differentiation of the inverse, 79
improper integral, 116
inactive function mode, 3
indeterminate form, 11, 94, 105
inequalities: a graphical approach, 84
integral
    as a function, 80
    CALC menu, 51
    constant of integration, 52
    fnInt( ) command, 51, 80
    Fundamental Theorem of Calculus, 51
    graphical, 51
    improper, 116
    numerical techniques, 55
    Simpson's Rule, 56
    sum(seq( )) command, 50, 57
    variable, 52
integral
    fnInt( ) command, 196
integral decimal approximations, 55
integral test, 116
integral vs. derivative, 53
integral-definition as a Riemann sum, 45, 50
Integration by Substitution Theorem, 54
intersection of planes, 163
interval of convergence, 121
inverse function, 74
inverse function program, 75
inverse function values, 75
inverse hyperbolic functions, 93

inverse trigonometric functions, 85

L'Hopital's Rule, 94, 105, 106
L'Hopital's Rule for a sequential limit,
    106
large window problems, 31, 139
Law of Cosines, 40
limit(s)
    a warning, 121
    complications, 13, 94, 95, 105–107,
        114
    formal definition, 13
    graphical, 10, 94, 95
    indeterminate form, 11, 105
    speeding things up, 12, 106
    table of values, 11, 23, 105, 106
line in space (parametrically), 175
Lissajous curves, 134
list(s)
    deleting from memory, 48, 164
    elements of, 76
    L-keys, 47, 153
    LIST menu, 47
    listable property, 48, 50
    maximum size, 48
    names, 47
    notation, 47
    operations on, 48
    removal from view screen, 48
    Riemann sums, 50
    seq( ) command, 48, 76
    SetUpEditor, 48
    special keyboard types, 47
    sum(seq( )) command, 50, 57
    sum( ) command, 50
    the L prefix, 47
    view screen editing, 48
    viewing screen, 47, 77
ln(x): a definition, 80
locally linear graphs, 21
logarithmic function, 80
logarithms: other bases, 84

Maclaurin series, 102
matrices, 77
MATRX menu, 77

maximizing an angle, 87
maximizing application, 8
maximizing area, 9
maximizing graphically, 4
Mean Value Theorem, 35
Mean Value Theorem for Integrals, 55
MEM menu, 3, 164
minimizing, 87
minimizing distance, 41, 137
minimizing graphically, 5
motion in space, 179
MTRX EDIT submenu, 78

natural logarithm, 80
nDer( ) (Ti-86): a warning, 24
nDer( ) command(Ti-86), 24
nDer( ) vs. der1( ) (Ti-86), 24
nDeriv( ) command, 24, 25
    a warning, 24
    not listable, 174
    on vectors, 174
    repeated applications, 177, 187
Newton's Law of Cooling, 91
Newton's Method: failure, 39
Newton's Method: geometry, 37, 39
numerical integration, 55

order of growth, 33
orthogonal frame, 185
overflow complications, 104, 105, 114

p-series, 113
parametric curve(s)
    arc length, 133
    function values, 132
    in the plane, 131
    intersection points, 135
    Lissajous, 134
    mode, 131
    planar graph, 132
    plotting $x = g(y)$, 133
    plotting $y = f(x)$, 133, 137
    plotting several:different intervals,
        132
    plotting two types, 133
    slope, 133, 136

　　　　symmetry, 134
　　　　trace cursor, 134, 137
　　　　zooming in, 135
parametric curve(s):arc length, 137
parametric vs. polar, 138, 139
piecewise functions, 7
plane, 162
plane intesections, 163
plotting
　　　　approximations, 124
　　　　asymptotic behavior, 33
　　　　complete graph?, 16, 26
　　　　concavity, 34
　　　　functions, 3
　　　　increasing/decreasing, 34
　　　　inverse information, 74
　　　　long term behavior, 33
　　　　order of growth, 33
　　　　parametric curves, 132, 133
　　　　polar graphs, 138
　　　　polynomials, 31
　　　　sharp corner, 16, 23
　　　　slant asymptote, 33
　　　　styles, 8, 25, 41
　　　　surprises, 16, 23, 26
　　　　Taylor polynomials, 123
　　　　Taylor's remainder, 125
plotting $x = g(y)$ parametrically, 133
plotting $y = f(x)$ parametrically, 133, 137
polar coordinates, 138
　　　　angle, 142
　　　　arc length, 139, 143
　　　　area, 139, 140
　　　　complete graph?, 141
　　　　conic sections, 143
　　　　function values, 140
　　　　intersection points, 140, 142
　　　　mode(coordinate), 140
　　　　mode(graphing), 139
　　　　nonuniqueness of points, 140
　　　　plotting, 138
　　　　plotting complications, 139, 140
　　　　slope, 139
　　　　trace cursor, 140, 142

polar vs. parametric, 138, 139
polynomials-long, 102
positive term series tests, 113
power series, 102, 121
principal unit normal vector, 185, 186
program(s), 108, 111, 148, 149, 158, 167
　　　　CATALOG menu, 76
　　　　CONTROL menu, 75
　　　　editing, 78
　　　　INPUT/OUTPUT menu, 75
　　　　names, 75
projection program, 167
projection vector, 161, 163

radius of convergence, 121
radius of curvature, 188
Ralph, 66, 169, 170, 181, 190, 191
rate of change, 69
ratio test, 114, 121
rational form, 161
recovering old statements, 38, 39, 50, 58, 67, 124, 141, 155, 167, 176, 177
recursively defined sequence, 103, 107
　　　　complications, 104
　　　　finance, 107
　　　　slow, 107
Riemann sums, 50
root test, 119, 121
roots of large polynomials, 36
round off errors, 13

scientific notation, 10
screens: making space, 35, 135
Seq mode, 103
seq( ) command, 48
sequence, *see* list(s)
　　　　a recursive type, 103
　　　　a warning, 121
　　　　as a function, 102
　　　　as a list, 103
　　　　complications, 106, 107
　　　　convergence/divergence, 105
　　　　definition, 102
　　　　fast program, 108

graphing complications, 103
increasing, decreasing, 104
L'Hopital's Rule, 105, 106
limit, 103, 104
overflow problems, 104, 105
recursive, 103, 107
    complications, 104
    finance, 107
    Finonacci, 109
    slow, 107
surprises, 105
table of values, 103
the Lprefix, 103
view screen, 108
series, *see* list(s)
    a definition, 109
    a warning, 121
    alternating test, 117, 119
    approximation, 115
    approximation error term, 116, 117
    as a sequence, 109, 112
    basic well known family, 113
    comparison tests, 113, 118
    complications, 114
    convergence absolute, 118
    convergence conditional, 118
    convergence fast, 118
    convergence/divergence, 113, 114, 116, 117
    convergence/divergence/fast/slow, 115
    decreasing, 116, 119
    error term, 115–118
    evaluation limitations, 110
    fast evaluation, 111
    geometric, 113
    harmonic, 113, 115
    integral test, 116
    interval of convergence, 121
    Maclaurin, 102
    overflow problems, 114
    p-series, 113
    positive term tests, 113
    power, 102, 121
    program, 111
    program (big sum), 111
    radius of convergence, 121
    ratio test, 114, 121
    root test, 119, 121
    sum(seq( )) command, 110
    table of values, 110
    Taylor, 102
    Taylor remainder, 124
    viewing screen, 111
    which test to use?, 115, 118, 119
SetLE (Ti-86), 48
SetUpEditor, 48, 176
sharp corner, 23
Simpson's Rule, 56
Simpson's Rule error term, 57
sine addition formula, 136
slant asymptote, 33
slope: computations, 22
slope: graphical, 22
smooth curve, 185
solving equations, *see* equation(s)
solving equations graphically, 1
special keyboard types, 153
speed, 179
standard window, 26, 34
STAT menu, 47, 77, 108, 111
storing to a name
    lists, 47, 79
    points, 162
    real numbers, 62, 67
    vector antiderivative functions, 174, 179
    vector derivative functions, 174, 177
    vector functions, 155, 173, 175
    vectors, 156, 159
sum(seq( )) command, 50
sum( ) command, 50

table of values, 9, 11, 42, 53, 110
tangent line, 26, 83
Taylor series, 102
Taylor's remainder, 124
TNB frame, 185
trace key, 4
transcendental functions, 73

unit normal vector, 185, 186
unit tangent vector, 186
unit vector, 154, 182

values of functions, 7
VARS menu, 7, 8, 12
vector valued function, 155, 173–175
    automatic updating, 175, 177
    table of values, 175
vector(s), *see* list(s)
    projection program, 167
    acceleration, 176, 179
    accessing components, 152
    addition, 153, 159
    algebraic properties, 159, 167
    as a list, 151
    as points, 162, 163
    binormal, 185, 187
    constant of integration, 178, 182
    creating unit length, 154
    cross product, 157, 158, 160, 162, 163, 177
    curvature, 186, 188
    deleting from memory, 164, 186
    differentiation, 177
    distance between lines, 162
    distance: point-line, 166
    distance: point-plane, 163
    dot product, 157, 158, 161, 162, 178
    equation of a plane, 162
    function, 155
    function values, 175
    L-keys, 152
    length, 166
    length (norm), 154, 161, 188
    lines, 166
    listable property, 153, 154
    normal, 177
    orthogonal, 161, 162
    orthogonal frame, 185
    plane intersections, 163
    position function, 182
    principal normal, 187
    program, 158, 167
    projection, 161, 163

projection program, 167
rational form, 161
scalar multiplication, 153
showing equality, 159
speed, 179
subtraction, 153
sum( ) command, 154
tangent, 175
TNB frame, 185
unit, 154, 182
unit normal, 185, 186
unit speed, 184
unit tangent, 184, 186, 187
velocity, 176, 179, 183
view screen, 176
VECTR menu (Ti-86), 151
velocity vector, 176, 179, 183
volume by cross-sections, 64
volume of a solid of revolution, 63

WINDOW menu, 4

Y= menu, 3

ZOOM menu, 4
ZoomFit, 4
ZSquare, 137
ZStandard, 26, 34